INTRODUCTION TO SOLAR AND INVERTER SYSTEM

A TRAINING MANUAL FOR INSTALLERS AND OPERATORS OF SOLAR AND INVERTER SYSTEM

O. E. ORESOTU

CHAPTER ONE — 4
1.1 HISTORY OF SOLAR AND INVERTER SYSTEM — 4

1.2 ADVANTAGES OF PHOTOVOLTAIC SYSTEM — 26

1.3 DISADVANTAGES OF PHOTOVOLTAIC SYSTEM — 29

1.4. COMPONENTS OF A PHOTOVOLTAIC SYSTEM — 30

CHAPTER TWO — 31
2.1 SOLAR RADIATION — 31
2.2 THE SOLAR GEOMETRY — 32
2.3 TYPES OF SOLAR RADIATION — 37

CHAPTER THREE — 38
3.1 PRINCIPLES OF PHOTOVOLTAIC SYSTEM — 38
3.2 COMPONENTS OF A PHOTOVOLTAIC SYSTEM — 39
3.3. TYPES OF PHOTOVOLTAIC SYSTEM OR SOLAR PANEL — 43
3.4. SOLAR PANEL WIRING — 49
3.5. PARAMETERS OF A SOLAR PANEL — 52
3.6. LOAD RESISTANCE — 55
3.7. EFFECT OF SOLAR LUMINANCE OR INTENSITY — 57
3.8. CELL TEMPERATURE — 58
3.9. SHADING — 58
3.10. DIODES — 62
3.11. BLOCKING DIODES — 62
3.12. BYPASS DIODES — 63

CHAPTER FOUR — 64
4.1. WORKING PRINCIPLE OF A LEAD ACID BATTERY — 64
4.2. TYPES OF BATTERIES AND THEIR OPERATION — 65
4.3. DAYS OF AUTONOMY — 73
4.4. BATTERY CAPACITY — 77
4.5. THE RATE AND DEPTH OF DISCHARGE — 78
4.6. BATTERY SAFETY — 80
4.7. BATTERY WIRING CONFIGURATION — 82

CHAPTER FIVE — 89
5.1. CHARGE CONTROLLER — 89
5.2. TYPES OF SOLAR CHARGE CONTROLLER — 90
5.3. CONTROLLER FEATURES — 95
5.4. TROUBLE SHOOTING A CHARGE CONTROLLER — 97
5.5 TIPS ON CHARGE CONTROLLERS — 100

CHAPTER SIX — 106
6.1. INVERTER — 106
6.2 CLASSIFICATION OF AN INVERTER — 109
6.3 INVERTER FEATURES — 117
6.4. SPECIFYING AN INVERTER — 118

CHAPTER SEVEN — 127
7.1. SIZING PHOTOVOLTAIC SYSTEM — 127
7.2. FACTOR OF DESIGN — 127
7.3. ELECTRIC LOAD ESTIMATION — 133
7.4. BATTERY SIZING — 142
7.5. SIZING YOUR SOLAR PANEL — 144
7.6. CONTROLLER SPECIFICATIONS — 151

CHAPTETR EIGHT — 160
8.1 PHOTOVOLTAIC SYSTEM APPLICATION — 160

CHAPTAER NINE — 167
9.1. SOLAR PANEL AND BATTERY MAINTENANCE — 167
9.2. TROUBLESHOOTING AN INVERTER — 177

CHAPTER ONE

1.1 History of solar and inverter system

Long before the first Earth day was celebrated on April 22, 1970, generating awareness about the environment and support for environmental protection, scientists were making the first discoveries in solar energy. It all began with ***Edmond Becquerel***, a young physicist working in France, in 1839.

Figure 1.1 Edmond Becquerel

He observed and discovered the photovoltaic effect, as a process that produces a voltage or electric current when exposed to light or radiant energy (sun).

Figure 1.2 Heinrich Rudolf Hertz

In 1887 **Heinrich Rudolf Hertz** was a German physicist, he observed the photoelectric effect, the production and the reception of electromagnetic waves. When light falls on a metal surface, some electrons near the surface absorb enough energy from the incident radiation to overcome the attraction of the positive ions in the material of the surface. After gaining sufficient energy from the incident light, the electrons escape from the surface of the metal into the surrounding space.

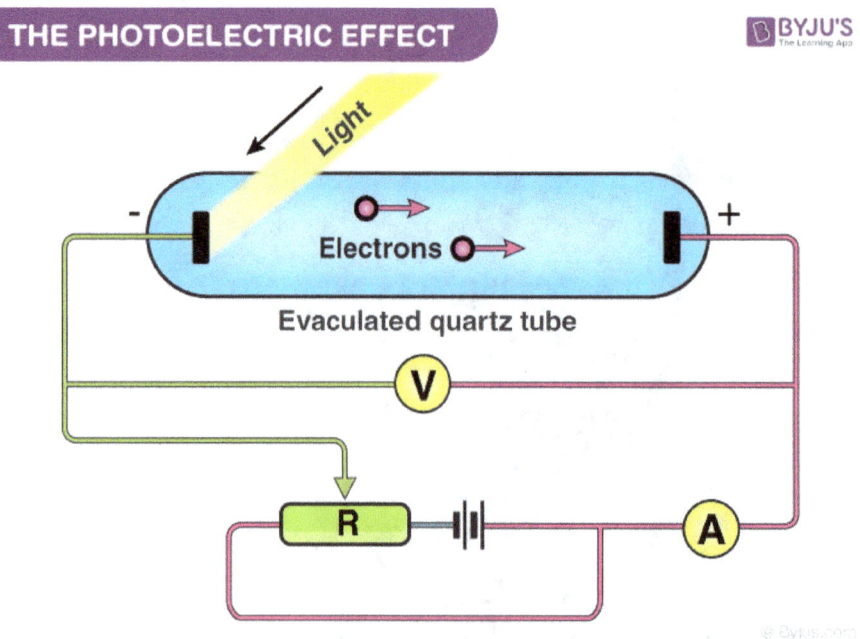

Figure 1.3 The photoelectric effect

From the image above, **Heinrich Rudolf Hertz** observed that when ultraviolet light shines on two metal electrodes with a voltage applied across them, the light changes the voltage at which sparking takes place. This relationship between light and electricity is known as "photovoltaic effects".

Twenty seven years later, a French mathematician *Augustin Mouchot* was inspired by the physicist's work.

Figure 1.4 Augustin Mouchot

He began registering patents (*Patent is a government authority or licensed conferring a right or title for a set period, especially the sole right to exclude others from making, using or selling an invention. A patent can also be an exclusive right granted for an invention, which is a product or a process that provides, in general, a new way of doing something, or offers a new technical solution to a problem*) for solar powered engines in the 1860s. From France to the U.S. inventors were inspired by the patents of the mathematician and filed for patents on solar powered devices as early as 1888.

Figure 1.5 Charles Fritts

Take a step back to 1838 when New York inventor **Charles Fritts** created the first solar cell by coating selenium with a thin layer of gold. Fritts **"reported that the selenium module produced a current that is continuous, constant, and of considerable force."** This cell achieved an energy conversion rate of 1 to 2 percent. Most modern solar cells work at an efficiency of 15 to 20 percent. So, Fritts created what was a low impact solar cell, but still, it was the beginning of photovoltaic solar panel innovation in America.

Figure 1.6 Alessandro Volta　　　　　Figure 1.7 First battery

The photovoltaic solar panel was named after Italian physicist, chemist and pioneer of electricity and power, *Alessandro Volta*, photovoltaic is the more technical term for turning light energy into electricity, and used interchangeably with the term photoelectric.

Figure 1.8 *Alessandro Volta, presents the first battery called the "voltaic pie" in an experiment.*

Alessandro Volta invented the first battery, which is known as "voltaic pie" today called dry cell battery, but it was non-rechargeable.

Figure 1.9 Gaston Planté First Rechargeable Battery

On the other hand, **Gaston Planté** was a French physicist who invented the lead acid battery in 1859. This type of battery was developed as the first rechargeable electric battery which is commonly used today.

Figure 1.10 Philipp Eduard Anton Von Lenard

In 1902 **Philipp Eduard Anton Von Lenard**, a Hungarian born German physicist observed the variation in electron energy with light frequency. He demonstrated that electrically charged particles are liberated from a metal surface when it is illuminated (exposed to sun ray or light) and that these particles are identical to electrons.

Figure 1.11 Albert Einstein

In 1905 **Albert Einstein** was a German born theoretical physicist, widely acknowledged to be one of the greatest and most influential physicists of all time.

He mathematically describe how the photoelectric effect was caused by absorption of quanta of light. Einstein assumed that a photon would penetrate the material and transfer its energy to an electron. As the electron moved through the metal at high speed and finally emerged from the material, its kinetic energy would diminish by an amount ϕ called the work function (similar to the electronic work function), which represents the energy required for the electron to escape the metal. By conservation of energy, this reasoning led Einstein to the photoelectric equation;

$E_k = hf - \phi$

Where E_k is the maximum kinetic energy of the ejected electron.

With this photo electric equation Albert Einstein described the nature of light and the photoelectric effect on which photovoltaic technology is based, for which he later won a Nobel Prize in physics. The first photovoltaic module was built by Bell Laboratories in 1954.

Figure 1.12 First solar battery

It was called solar battery and was mostly just a curiosity as it was too expensive to gain widespread use. In the 1960s, the space industry began to make the first serious use of the technology to provide power aboard spacecraft.

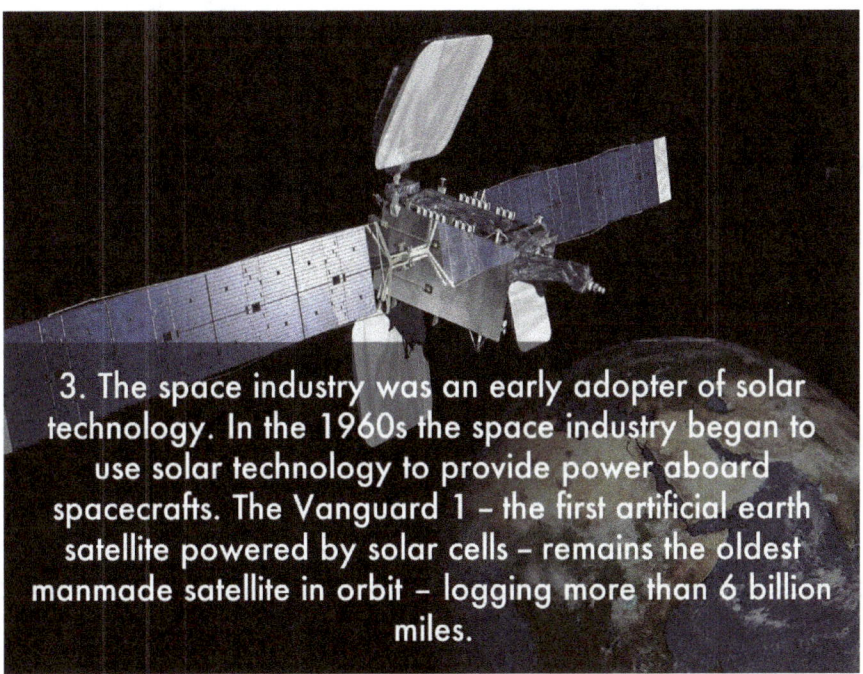

Figure 1.13 First solar application in space

Figure 1.14 Galatasaray stadium in Turkey as a non-space application of solar power

Through the space programs, the technology advanced, its reliability was established, and the cost began to decline. During the energy crisis in the 1970s, photovoltaic technology gained recognition as a source of power for non-space applications.

Figure 1.15 Enerjisa energy

In March 9, 2021 Galatasaray signed its first deal with **Enerjisa**.

Enerjisa energy the pioneer and leading Energy Company of the electricity distribution and retail sector in Turkey, has received a Guinness World Record titled *"most powerful solar power output from a sport stadium"* in terms of installed capacity and gone down in history. The installation has begun generating electricity.

Figure 1.16 Enerjisa energy receives the Guinness World Record

Enerjisa used a Goodwe's 120KW HT series inverters for this solar power plant with an installed capacity of 4.2MW on the roof of the Ali Sami Yen Sports Complex Nef Stadium in Galatasaray.

Figure 1.17 Galatasaray stadium as first in history

The rooftop power plant is to have 10,000 solar panels with a total area of 40,000 square meters. Its transmission capacity will be able to supply electricity to 2,000 households. The installation will beat the Estadio Nacional de Brasilia Mane Garrincha power plant, which produces 2.5MW per year.

Other non-space application of solar system is shown in the image below

Figure 1.18 Non-space applications of solar system

Also, the modern solar inverter, in its simplest form, is a **power converter**. It converts DC power produced by solar panels into AC power for the appliances we use in our homes. But the first power converters did the opposite. They took AC power and converted it to DC.

Figure 1.19 The rotary converter

Charles S. Bradley invented the rotary converter in 1888. At the time most appliances and machinery operated on DC power, but AC transmission was quickly becoming dominant. There was a pressing need to convert the transmitted AC power into DC that the appliances could use.

Figure 1.20 Charles S. Bradley, inventor of rotary converter.

Rotary converters and also motor generator sets did this job from the late nineteenth century until the mid-twentieth century. Now when we talk about power converters like this, that can convert AC to DC, we usually just call them **'rectifiers'**.

The first known use of the term **"inverter"** was in 1925 by **Engr. David Prince**, He published an article in the GE Review in which he wrote;

Figure 1.21 Engr. David Prince

"The author took the rectifier circuit and "inverted" it, turning in direct current at one end and drawing out alternating current at the other"

So, a solar inverter is called an inverter because it reverses, or 'inverts' a rectifier's operation.

In the 1950s inverters moved from being mechanical devices to ones with solid state circuits. This was made possible by the dawn of a new field of engineering called 'power electronics'.

Figure 1.22 Inverter circuit

Regular electronics were dominated by the humble transistor, which had insufficient voltage and current ratings for most inverter applications. Power electronics used a type of high power transistor called a thyristor or SCR (silicon-controlled rectifier). In 1953 German company Kaco manufactured the world's first thyristor inverter.

Figure 1.23 Kaco manufactured the world's first thyristor inverter

Years later Kaco would go on to produce the first transformer-less inverter. In 1999 a handful of "idealists" clambered onto the rooftops of homes in Baden-Württemberg to install solar PV systems. Accompanying them was the world's first transformer-less string solar inverter, the Kaco Blue Planet.

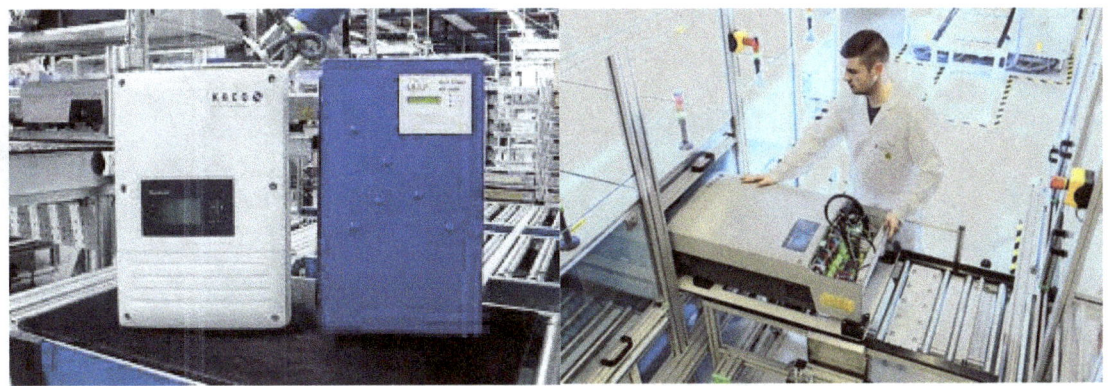

Figure 1.24 Kaco produced the world first transformer less inverter

The introduction of transformer-less inverters in 1999 brought with it many advantages being more efficient, lighter and cheaper to manufacture. The main drawback however was the lack of galvanic isolation between the DC and AC circuits, which could potentially allow the passage of dangerous DC faults to the AC side.

Galvanic isolation (provided by a transformer) separates the input and output supplies to a device so that energy flows through a field rather than via electrical connections. It enables power transfer between two circuits without being electrically connected.

Transformer-less solar inverters created more than a few headaches for electricity network providers and resulted in another rewrite for industry standards associations around the world.

In June 2008 Californian Company **Enphase** introduced their first micro inverter.

Figure 1.25 Enphase world first micro inverter

It was about the size of a paperback book and was going to flip the solar industry on its head. So far, that hasn't happened.

The concept is a great idea because it sits underneath each solar panel and allows each panel in an array to operate independently of one another. In the event of one or multiple panels reducing output due to shading or failure, the rest of the array will keep chugging along nicely. This is not the case with a single series string of solar panels feeding a string inverter.

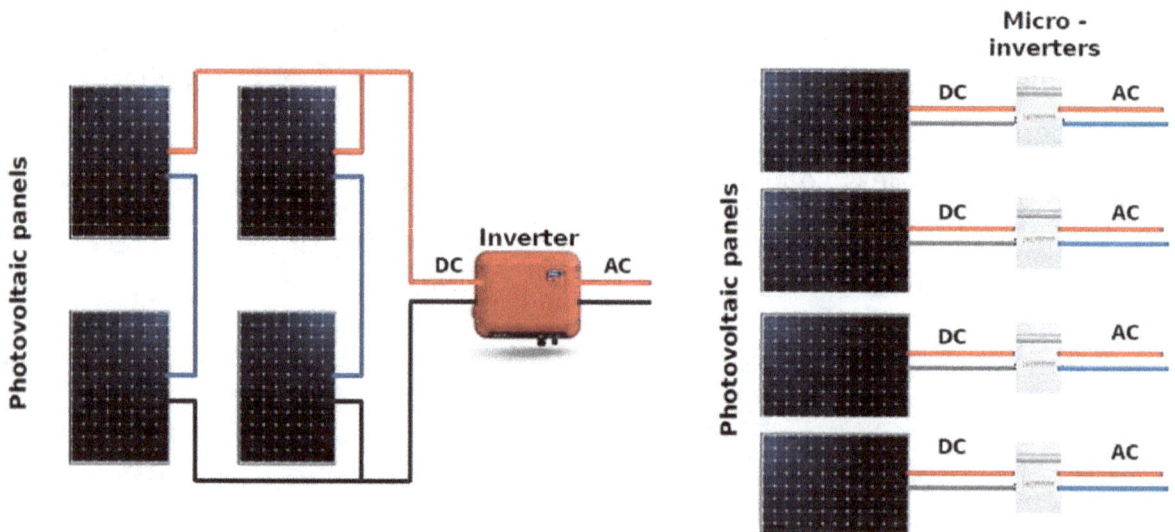

Figure 1.26 String inverter vs. Micro inverter

So, between string inverter and micro inverter which one will you prefer…?
No answers yet…!

Other advantages are numerous. They include panel level optimization and monitoring, safer due to lower DC voltage and no DC cabling in roof space, panels can be oriented in different directions, simpler to design, panel model agnostic, easier PV array expansion and much longer inverter warranties.

The reason this technology hasn't taken over the world is cost. They are expensive, complex to install and may have higher maintenance costs due to the amount of hardware on the roof.

Also, Hybrid inverters have been around for a while but are starting to gain some serious attention due to the increasing uptake of home battery systems.

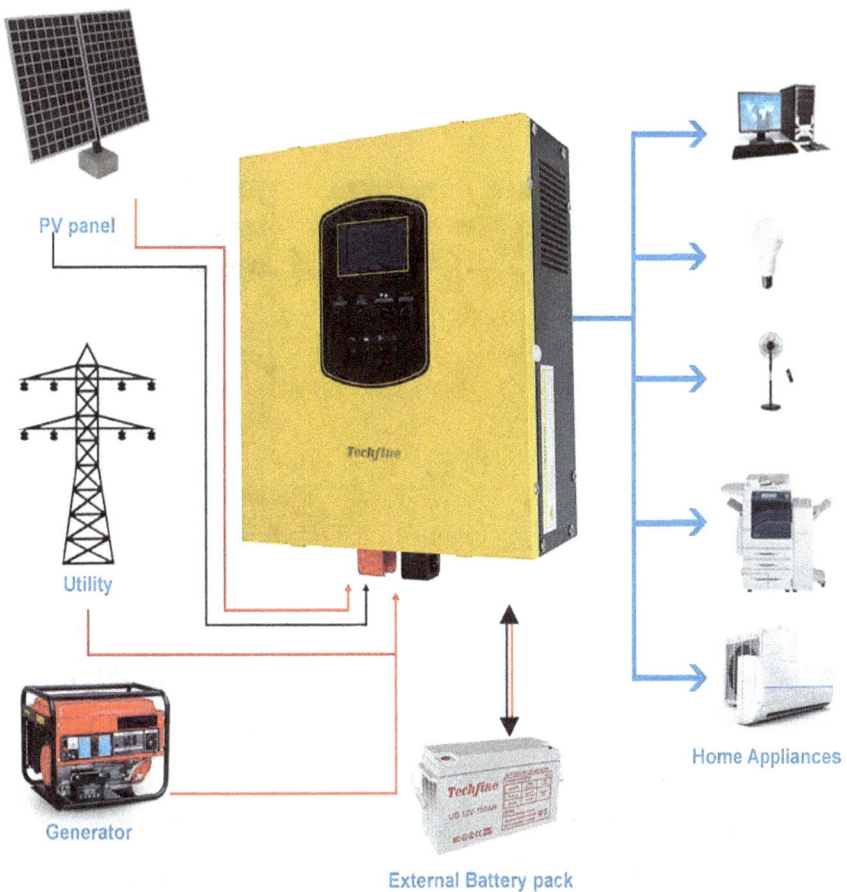

Figure 1.27 Hybrid inverter

These inverters are sophisticated, intelligent beasts that can simultaneously manage inputs from solar panels, a battery bank and utility grid using a technique called DC coupling (along with more than a few other bells and whistles).

The hybrid inverter's brain can decide where to draw energy from, when and how to do it. It can make decisions based on the price of electricity at any given time, and even draw energy from the grid to store in the battery if that's economical.

We're likely to see more of these inverters from now on due to the growing realization around the world of the importance of embracing technologies that will deliver zero carbon emissions.

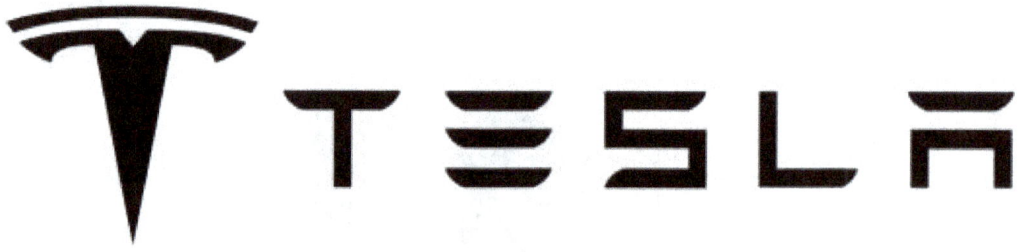

Figure 1.28 Tesla Energy

In 2019, American company called **Tesla** invented a bigger sophisticated beast called the tesla mega-pack. This is a solar and inverter system that can be used in a giga scale projects, it can be used for energy storage and transmission in large scales. The tesla mega-pack is a game changer in the power sector because it uses the power of the wind and sun to generate electricity in larger scales and can be transmitted in large scales for commercial and industrial uses. Tesla mega-pack can replace hydroelectric power plants, nuclear power plants and other sources of power generation.

Figure 1.29 Tesla mega-pack and inverter

Each Mega-packs comes from the factory fully assembled with up to 3 megawatt hour (MWhs) of storage and 1.5MW of inverter capacity, building on power-packs engineering with an AC interface and 60% increase in energy density to achieve significant cost and time savings compared to other battery systems and traditional fossil fuel power plants such as gas plants, nuclear plants etc.

Using mega-pack, tesla energy can deploy an emission free 250MW, 1GWh power plant in less than three months on a three-acre footprints which is four times faster than a traditional fossil fuel power plant of the same size.

Mega-pack can also be DC connected to solar, wind energy creating seamless renewable energy plants.

Figure 1.30 Tesla mega-pack and inverter powered by solar and wind energy

Mega-pack will act as a sustainable alternative to natural gas "peaker" power plants. Peaker power plants usually fire up whenever the local utility grid supply can't provide enough power to meet peak demand. Instead, a mega-pack installation can use stored solar or wind energy to support the grids peak load.

1.2 Advantages of photovoltaic system

Photovoltaic systems form a reasonable alternative to conventional power supply by generation systems based on conventional fossil fuel based technology e.g. diesel generator sets, Kerosene lamps, dry cell batteries.

1. **Clean and green energy source**
   The most important advantage of PV system is the clean and green energy it provides, because there is no fear or worry about the panels generating any harmful greenhouse gasses into the air such as sulfur dioxide (SO_2), nitrogen oxides (NO_X), particulate matter (PM), carbon dioxide (CO_2), mercury (Hg), and

other pollutants which are harmful for human and animal consumption and also may increase the effects of global warming.

2. Reliability

Even in harsh environmental conditions, photovoltaic systems have proven their reliability. PV arrays provide stable power supply in situations where continuous operation of other technologies is critical.

3. Free raw material

Another advantages is that you don't have to buy raw material, PV cells depends fully on the solar energy to produce the electricity needed to power your home or office, which is available in abundance every day as far as nature is concerned.

This energy when converted to electricity and stored in battery allows you save up in electricity costs once you start using the energy generated by the photovoltaic system.

4. No fuel used

Since no fuel source is required, there are no costs associated with purchasing, storing or transporting fuel. However, batteries need attention, and have to be replaced after a couple of years.

5. Durability

Crystalline photovoltaic cells has shown no degradation even after 25 years of use, while amorphous cells show degradation after some years. It is likely that future crystalline cells will last even longer. Crucial for long lifetime of PV panels is a reliable framing protecting the cells against humidity and dust.

6. Versatility

Solar PV cells can generate electricity anywhere. All it requires is sunlight, making it a useful energy source while going on camping trip, traveling, long car trips etc.

7. Significant impact on smart energy networks

Solar PV has an integral role in the smart energy networks, which work on distribution power generation (DPG). DPG are exceptionally environmentally friendly because it helps reduce the production of electricity at centralized power plants.

Besides, distributed power generation reduces the environmental impacts of a centralized power plants.

Also, distributed power generation (DPG) reduces any possible loss of energy during the transmission and distribution of electricity in the power system.

8. Independence

PV systems can be operated in remote areas independently of national or local grids. However, the energy service provided is rather limited, compared with grid connection or diesel generator sets, due to the high investment costs, urging the system size as small as possible.

9. Low maintenance cost

Solar PV cells are known for their low maintenance and operating cost compared to other renewable energy system.

Also the cost of maintaining a nuclear plant, hydro-electric power plant, generators, gas plants etc. can't be compared with the cost of maintaining solar plant.

10. Zero noise

Solar PV is perfect for urban areas and residential applications because it doesn't produce noise. When running a hotel or hospital business where zero noise is needed for the comforts of clients and patients, PV system can be used to generate power compare to emergency generators that produces noise.

11. Easy to install

Last but not the least, you can easily install residential solar plants on rooftop or just on the ground without interfering with your lifestyle.

1.3 Disadvantages of photovoltaic system

Everything that has an advantages also have a disadvantages. The disadvantages of a solar and inverter system includes the following.

1. Intermittency problems

Like all other renewable energy sources, solar energy PV cells have intermittency problems. It means it's not continuously available for converting into electricity like during night-time and during cloudy weather. So PV cell will probably be incapable of meeting an electric power systems demand.

2. It is expensive

PV cells require an additional investment in inverter and storage batteries. Inverter convert direct electricity to alternating electricity to use on your power network. The storage batteries prove helpful in providing a continuous power of electricity on-grid connections. This addition investment can, however, provide a solution for the PV cells intermittency problems.

3. A large area of land is used per square meter

The large areas of land used for land-mounted PV panels installation remains committed to the purpose. That's why it is important to wisely select a spot for solar energy system, and why many people install it upon their roof.

4. Easily damaged

While solar PV needs no maintenance or operating cost, its fragility means it's easily damaged. There is solution here in the form of additional insurance coverage to safeguard your investment.

1.4. Components of a photovoltaic system

Photovoltaic systems consist of several components;

1. **Direct current loads:** DC loads are appliances such as light bulbs, radio, TV, fans, motors, refrigerators, etc. powered by direct current are called DC loads.
2. **Solar array:** A PV array is a combination of more than one solar panel connected either in series, parallel or combination of both in conjunction to produce electricity
3. **Alternating current loads:** AC loads are appliances such as light bulbs, radio, TV, fans, motors, refrigerators, etc. powered by alternating current are called AC loads

4. **Charge controller:** charge controller is used to regulate battery charging and discharging. Charge controllers can be categorized into five categories.
    a) Shunt controller
    b) Series regulators
    c) Pulse Width Modulation (PWM)
    d) Maximum Power Point Tracking (MPPT)
    e) Diversion load Controllers
5. **Battery**: Battery is a storage device that stores electrical energy (DC) in the form of chemical energy. There are different types of batteries.
    a) Dry cell battery
    b) Lead acid battery
    c) Lithium ion battery
6. **Solar panel**: Solar panels are those devices which are used to absorb the sun's rays and convert them into electricity or heat. Solar panel can give 12VDC output voltage in normal cases, but bigger PV panels provides DC currents at 33.5VDC and also 44.3VDC.

NOTE: *Solar panels are rated in watts (W). A 12VDC panel can have a maximum power of 100W, 33.5VDC solar panel can have a maximum power of 350W and a 44.3VDC solar panel can have a maximum power of 540W as the case may be.*

CHAPTER TWO

2.1 Solar radiation

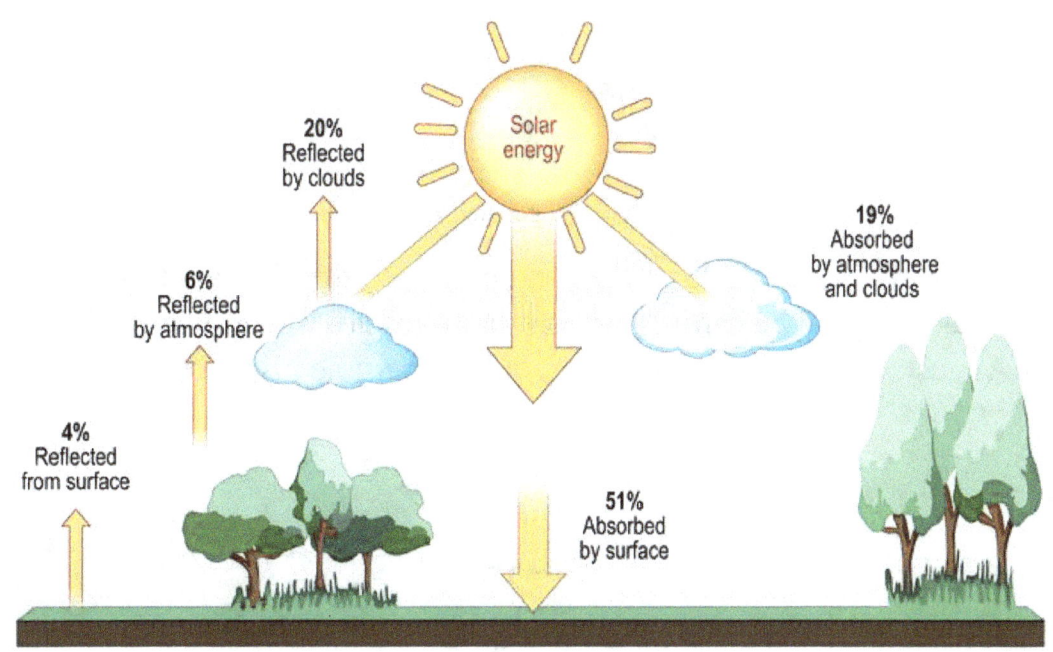

Figure 2.1 Solar radiation

The amount of solar radiation striking a surface at a particular time and place is known as insolation. When insolation is described as power, it is expressed as number of watts per square meter and usually presented as an average daily value for each month. On a clear day, the total insolation striking the earth is about 1,000 watts per square meter. However, many factors determine how much sunlight will be available at a given site, including atmospheric conditions, the earth's position in relation to the sun, and obstructions at the site.

2.2 The solar geometry

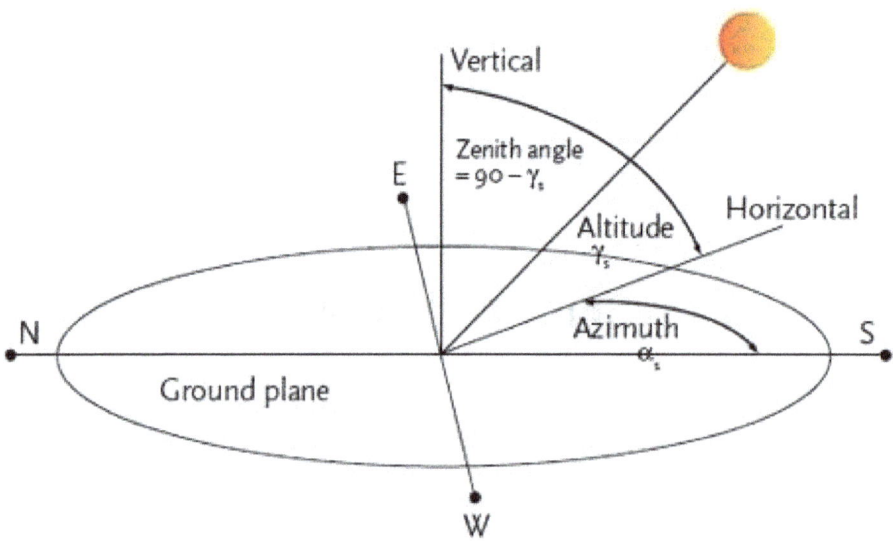

Figure 2.2 Solar geometry.

The earth's distance from the sun and the earth's tilt affect the amount of available solar energy. The earth's northern latitudes are tilted towards the sun from June to August, which brings summer to the northern hemisphere.

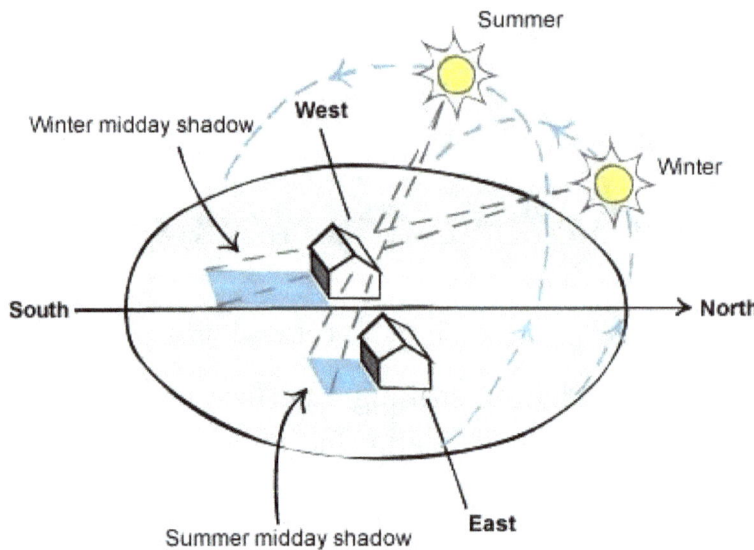

Figure 2.3 Sun position during summer and winter

The longer summer days and the more favorable tilt of the earth's axis create significantly more available energy on a summer day than on a winter day. In the northern hemisphere, where the sun is predominantly in the southern sky, photovoltaic modules should point towards the southern sky to collect solar energy.

Designers should optimize solar collection by positioning the array to take full advantage of the maximum amount of sunlight available at a particular location.

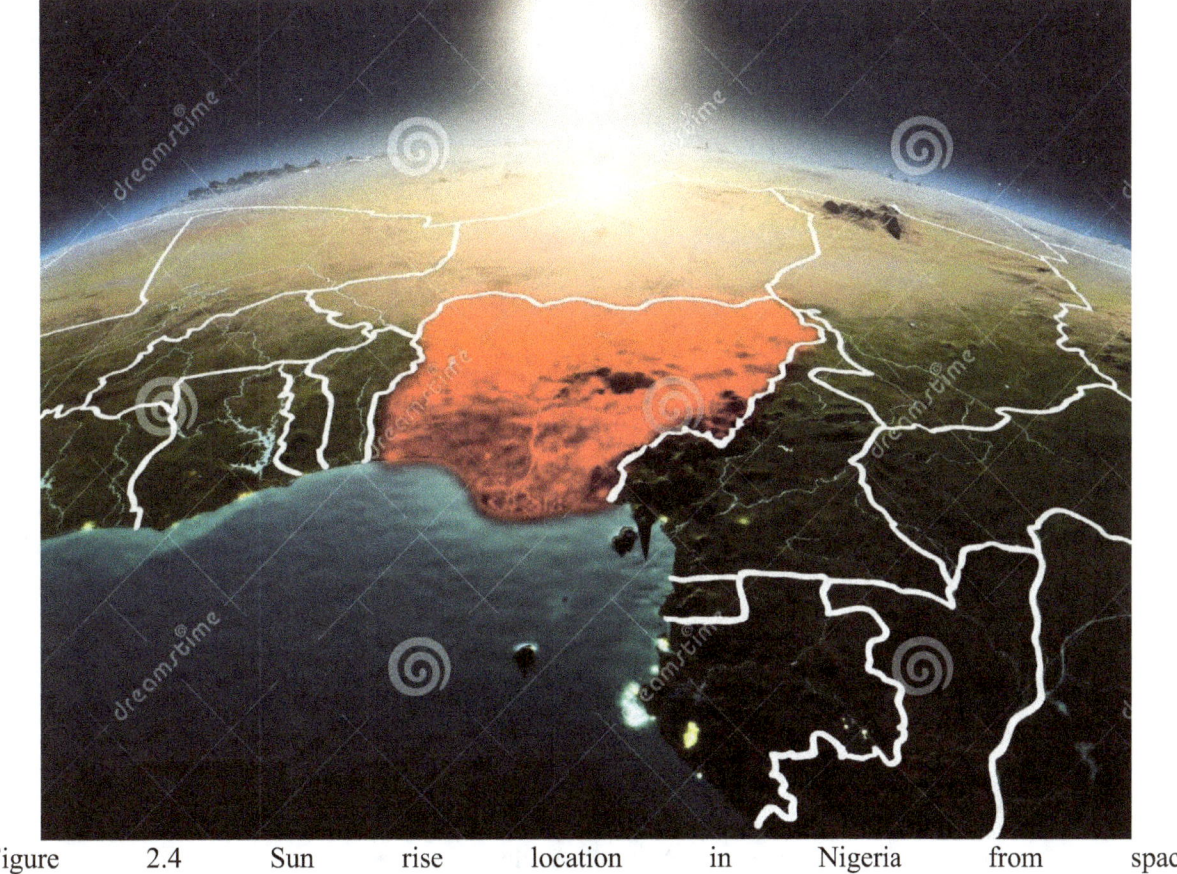

Figure 2.4 Sun rise location in Nigeria from space

Fortunately, the sun's path across the sky is orderly and predictable. The site's latitude (the distance north or south of the earth's equator) determines whether the sun appears to travel in the northern or southern sky. For example Lagos, Nigeria is located at approximately 9.0820°N, 8.6753°E latitude, and the sun moves across the southern sky. At midday, the sun is exactly true south.

2.2.1. Orientation

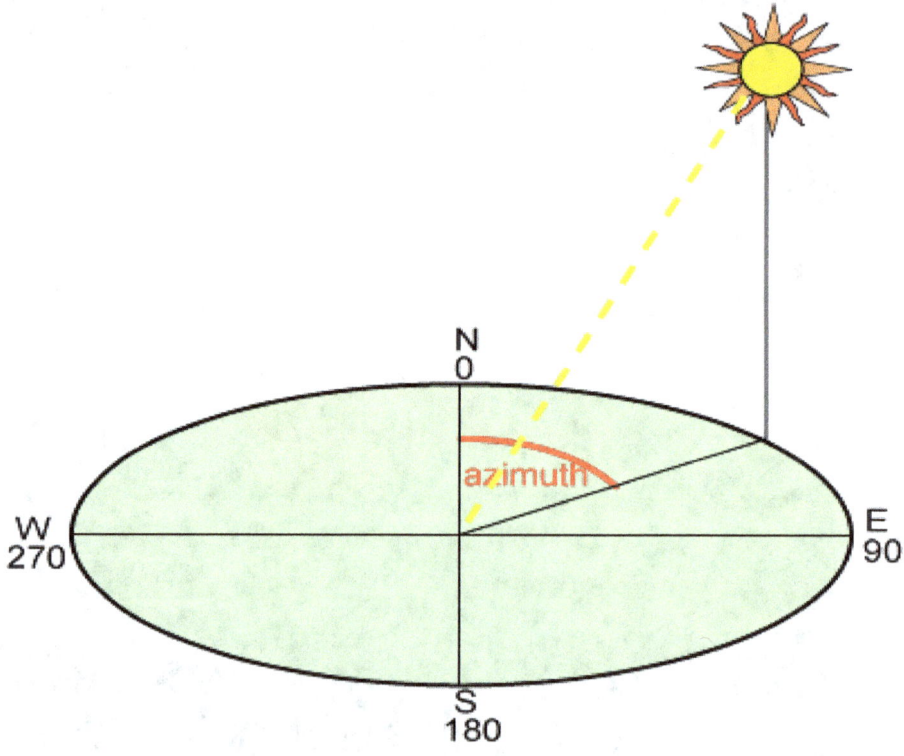

Figure 2.5 Azimuth Angle

The sun's apparent location east and west of true south is called azimuth, which is measured in degrees east or west of true south. Since there are 360 degrees in a circle and 24 hours in a day, the sun appears to move 15 degrees in azimuth each hour (360 degrees divided by 24 hours). Daily performance will be optimized, if fixed modules are faced true south. An array that deviates 15 degrees from true south will collect 90 percent of the sun's available energy on an average daily basis.

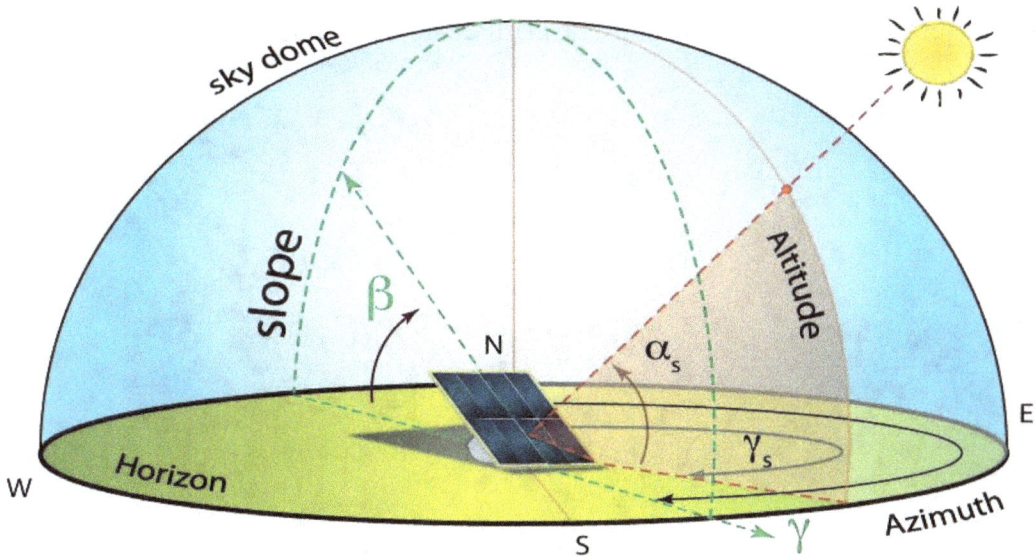

Figure 2.6 *Azimuth and altitude for all northern latitudes*

2.2.2 Tilt angle

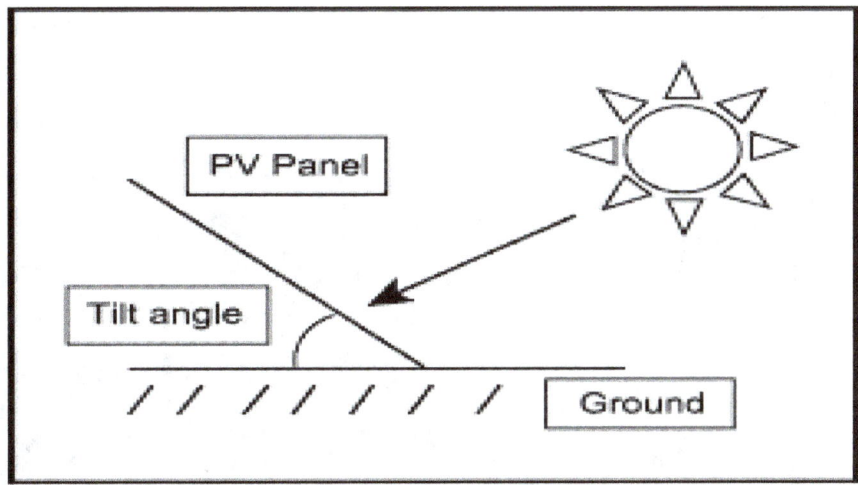

Figure 2.7 Tilt angle

The sun's height above the horizon is called altitude, which is measured in degrees above the horizon. When the sun appears to be just rising or just setting, its altitude is 0 degrees. When the sun is true south in the sky at 0 degrees azimuth, it will be at its highest altitude for that day. This time is called solar noon. A location's latitude determines how high the sun appears above the horizon at solar noon throughout the year. As a result of the earth's orbit around the sun with a tilted axis, the sun is at different altitudes above the horizon at solar noon throughout the year. The highest

Average insolation will fall on a collector with a tilt angle equal to the latitude.

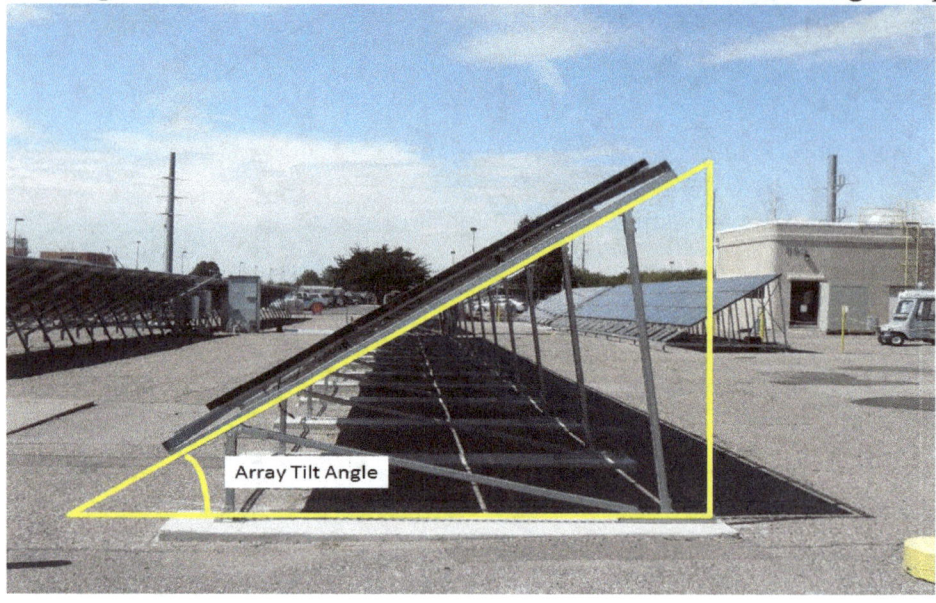

Figure 2.8 Tilt angle of a solar array

Adjusting the tilt angle of the PV array seasonally can increase power production significantly for year-round loads. Photovoltaic arrays work best when the sun's rays strike perpendicular (90 degrees) to the cells. When the cells are directly facing the sun in both azimuth and altitude, the angle of incidence is "normal".

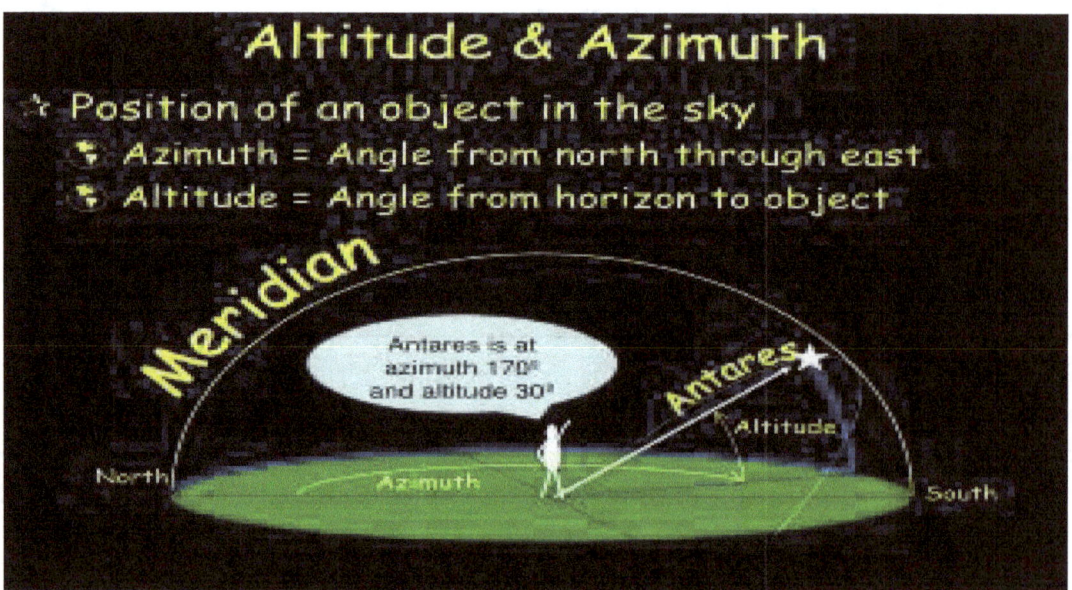

Figure 2.9 Azimuth And Altitude

2.3 Types of solar radiation

Solar radiation received at the earth's surface can be divided into two types.

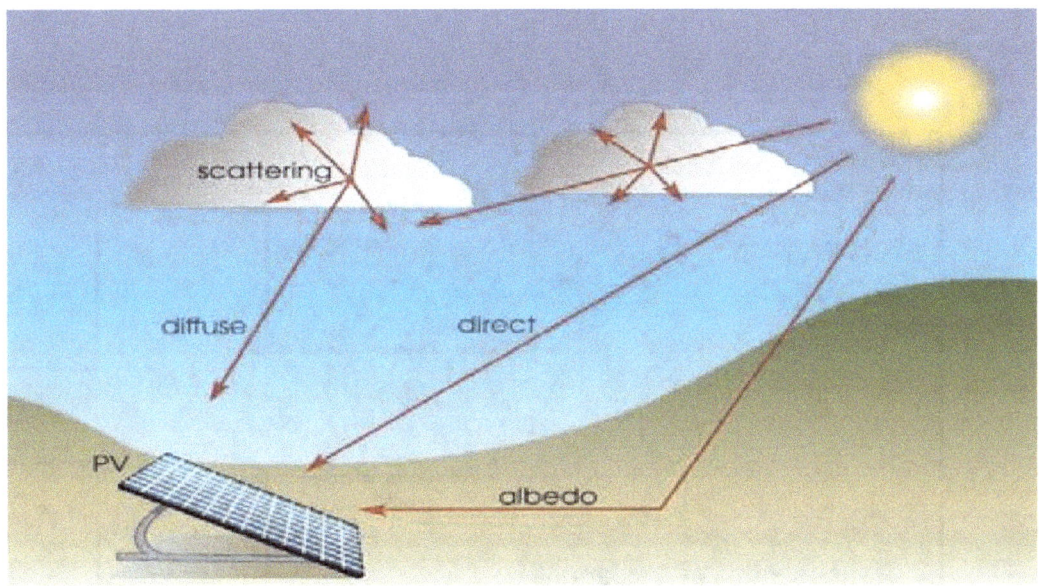

Figure 2.10 Solar Radiation

2.3.1 Direct beam solar radiation

Sunlight that hits the earth's surface directly is called direct radiation. In contrast to diffuse radiation, the light hits the earth's surface without any obstacles, e.g. clouds, trees, etc. Direct radiation can therefore be defined as any radiation that hits an object by the shortest route, from which it is then partially transmitted as diffuse radiation. PV plants absorb parts of the solar energy and convert it into electrical energy. However, large parts are also reflected and scattered as diffuse radiation. Together with the diffuse radiation, the direct radiation forms the global radiation.

2.3.2 Diffused solar radiation

Diffuse sky radiation is solar radiation reaching the Earth's surface after having been scattered from the direct solar beam by molecules or particulates in the atmosphere. Also called sky radiation, diffuse skylight, or just skylight, it is the reason for the color changes of the sky.

CHAPTER THREE

3.1 Principles of photovoltaic system

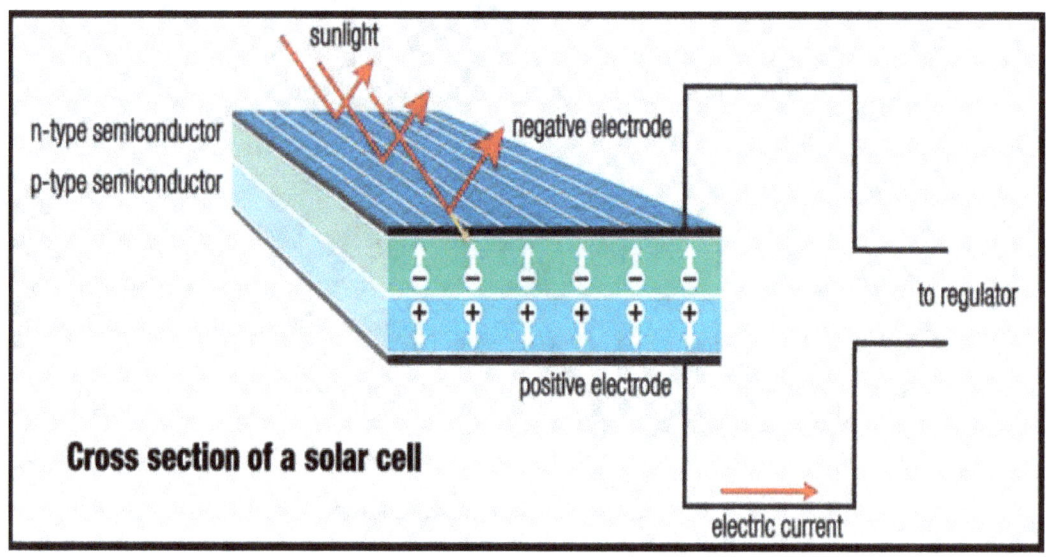

Figure 3.1 Working principle of a photovoltaic system

The Photovoltaic system consists of two layers of a semiconductor material which are the n-type semiconductor and the p-type semiconductor, usually silicon. **When sun rays or sunlight strikes these materials it creates an electric field across the layers causing electricity to flow**. The higher the intensity of the sun rays, the higher the flow of electricity and electricity produced.

The reason why Silicon is the most commonly used semiconductor for making solar cell. Is because Silicon carries millions of tiny atoms that have charged electrons. The most common design of solar panels today uses two different types of silicon – positively charged (P-type) and negatively charged (N-type). This is to create solar cell made up of two different layers, a layer of positively charged silicon and a layer of negatively charged silicon sandwiched together. To achieve this, small quantities of other elements is squeezed into the silicon layers. The silicon in the top layer is combined with phosphorus atoms which contain more electrons to create a negatively charged silicon (N-type) layer and the bottom layer gets a dose of boron, which contain less electrons to create a positively charged silicon (P-type) layer. When these two different layers of opposite charges are separated and placed side by side inside a solar cell it creates an electric field across the cell just like in a battery. This unique configuration enables a reaction that produces electricity when the silicon cells are exposed to sunlight.

3.2 Components of a photovoltaic system

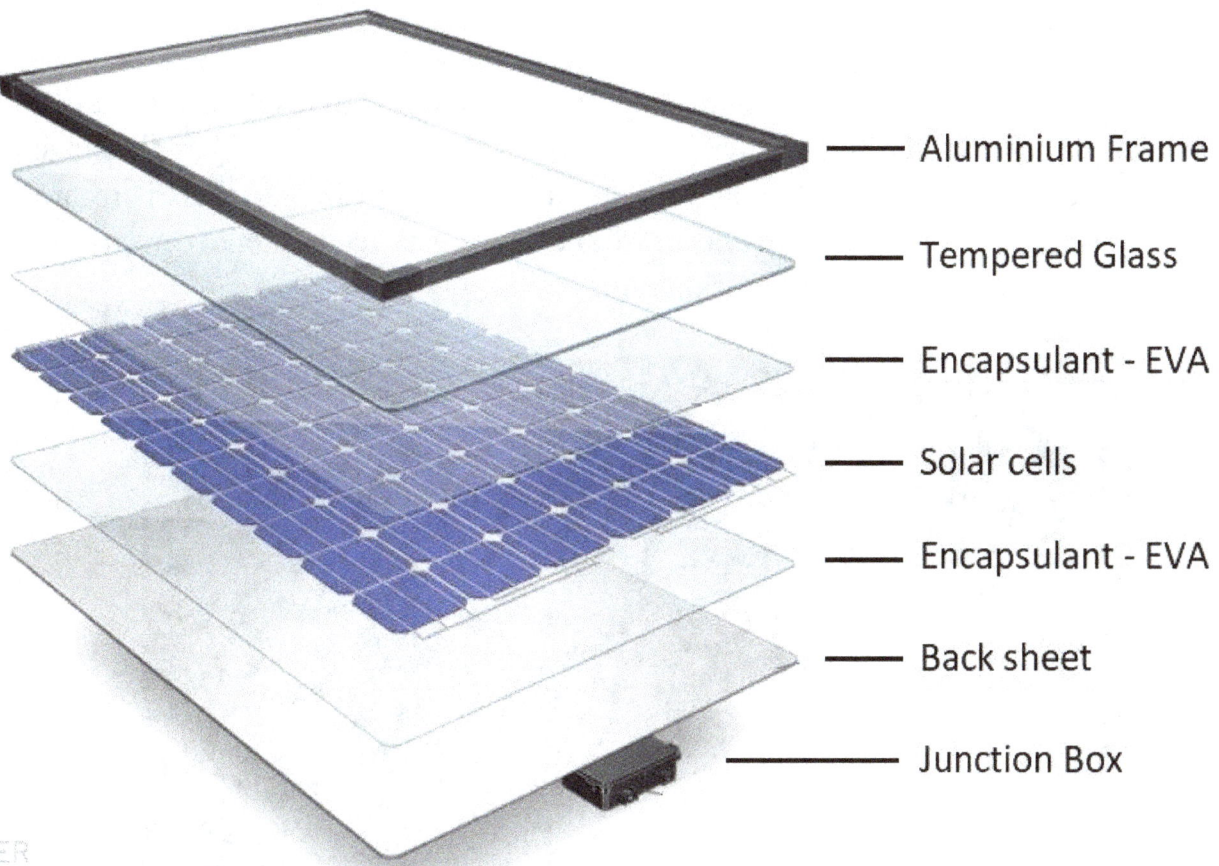

Figure 3.2 Components of a photovoltaic system

3.1.1 Aluminum frame

The aluminum frame is an important component as it provides structural strength to the panel. It is recommended to use a frame made of strong but lightweight material. It should be stiff and able to withstand extreme conditions like high wind and external forces. It generally comes in two makes - silver and anodised black.

Figure 3.3 Aluminum frame

3.2.2 Tempered solar glass

Tempered solar glass is another important component of a solar panel. It is the outer most layer on the solar panel and has to be strongly and solidly built and shiny for better performance of the panel. The main function of solar glass is to protect the solar cells from harsh weather, dirt and dust. It is recommended to use tempered glass with 3mm - 4mmm thickness.

Figure 3.4 Tempered solar glass

3.1.2 Encapsulated foil (Eva)

The EVA sheet or the *'ethylene vinyl acetate'* is a highly transparent (plastic) layer used to encapsulate the cells. It provides laminated layering on top of the cells to hold them together. It should be durable as well as tolerant to withstand extreme temperatures and humidity.

Figure 3.5 Encapsulated foil

3.1.3 Stringed solar cell

Solar cells are the building blocks of solar panels. Thousands of cells come together to form a solar panel. These Solar Cells are stringed together to make Solar Panels which involves soldering, encapsulating, mounting them on a metal frame, testing etc. The efficiency of a solar panel is directly proportionate to that of solar cells. The cost and efficiency of solar cells influence the overall performance of the solar panel. Solar Cell Efficiencies have improved continuously over the past decade.

Figure 3.6 Stringed solar cell

3.2.5 Back sheet

Figure 3.7 Black sheet

Back sheet is the rear-most layer of the panel providing both mechanical protection and electrical insulation. It is essentially a protective layer.

3.2.6 Junction box

A junction box is fixed at the backside of the panel. It is the central point where cables interconnect with the panels. The junction box has a diode which allow current to flow through it in only one direction.

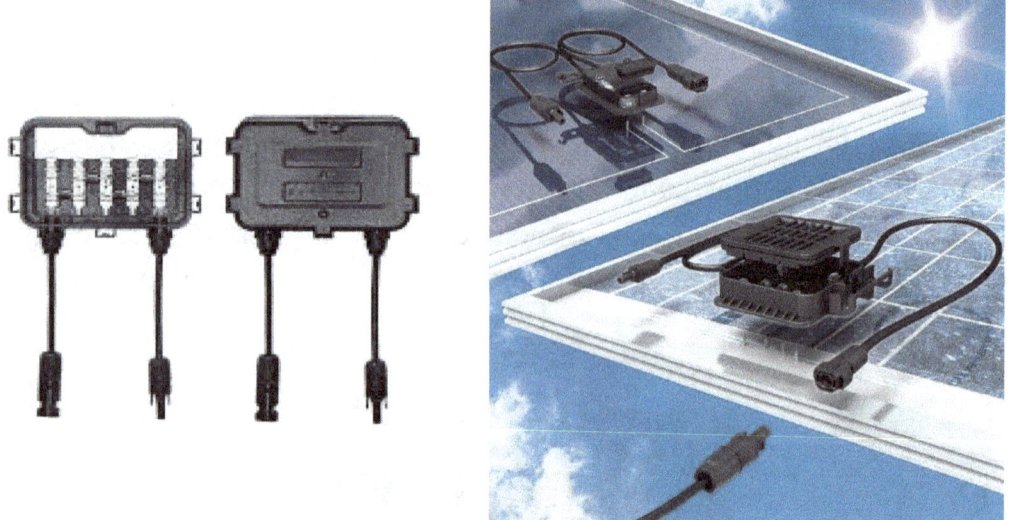

Figure 3.8 Junction box

ELECTRIFIED

3.2.7 Tabbing Wires

Tabbing wires are also called Interconnectors help solar panels connect with one another. These should be extremely weather-resistant and should enable secure connections.

Figure 3.8 Tabbing Wire

3.3. Types of photovoltaic system or solar panel

Basically there are three types of solar panel:
- Monocrystalline solar panel
- Polycrystalline solar panel
- Thin film solar panel.

3.2.7 **Monocrystalline solar panel**: Monocrystalline solar panels are also called single-crystalline silicon. Why? This is because they are made from growing a single crystal. Monocrystalline solar panels have an efficiency rate of between 15 to 20%.

They are the most efficient types of solar panels because their crystal framework is even. Physically, they appear in grid-like structures.

Figure 3.9 Monocrystalline solar panel

3.3.1.1 Advantages of monocrystalline

1. **Longevity**

   Monocrystalline solar panels are first generation solar technology and have been around a long time, providing evidence of their durability and longevity. The technology, installation, performance issues are all understood. Several of the early modules installed in the 1970's are still producing electricity today. Single crystal panels have even withstood the rigors of space travel.

   Some solar websites suggested that single crystalline solar panels can last up to 50 years. According to solar engineers I speak with even though this may be possible, there will be a slight drop off in efficiency of around 0.5% on average per year. So although this type of solar panels can last a long time, there will come a time when the lower efficiency makes it economically desirable to replace the panels especially as the efficiency of newer panels continues to increase.

   Note: Most performance warranties go for 25 years, but as long as the PV panel is kept clean it will continue to produce electricity.

2. **Efficiency**

   As already mentioned, PV panels made from monocrystalline solar cells are able to convert the highest amount of solar energy into electricity of any type of flat solar panel. Consequently, if your goal is to produce the most electricity from a specific area (e.g. on a roof) this type of panel should certainly be considered.

   Consequently, Monocrystalline panels are a great choice for urban settings or where space is limited. As a developer of PV rooftop installations in Germany, buying or leasing roof space is a significant cost of the whole project and so you want to be able to produce as much electricity you can from this valuable resource.

3. **More electricity**

   Besides producing more electricity per square meter of installed panels, thereby improving your cash flow (from either a reduction in your electrical bill or from the sale of the electricity or in some areas both), for those who are "going green" and are concerned about the environmental impact of solar panels, monocrystalline panels reduce the amount of electricity needed from local power plants, reducing the dependence on fossil fuels. The greater benefit is a reduction in the use of limited fuel sources and greenhouse gases being pumped into the environment.

4. **Greater heat resistance:** Like other types of solar panels, monocrystalline solar modules suffer a reduction in output once the temperature from the sunlight reaches around fifty degrees Celsius/ hundred and fifteen degrees Fahrenheit. Reductions of between twelve and fifteen percent can be expected. This loss of efficiency is lower than what is typically experienced by owners of PV panels made from polycrystalline cells.

3.3.1.2 Disadvantages of a monocrystalline solar panel
1. **Initial cost**

Because PV panels are made from single-cell silicon crystals, the process of making them is one of the most complex and costly ones around. Good silicon feedstock is expensive (although less in 2010 than it has been for a while) and the cost of making a single pure crystal is time-consuming and therefore costly, PV panels from monocrystalline solar cells generally cost more per panel than competing PV technologies.

However, instead of comparing costs on a per panel basis, or even on a per kW basis. I recommend you look at the investment on a per kWh basis over the expected life of the panel. Based on the analyses I've seen of various project proposals, monocrystalline solar panels are typically the most economical over the life of the installation.

2. Fragile

You should take into consideration that solar panels can be broken by tree branches or by objects carried by a strong wind. Generally, the solar panels are covered by a safety glass that helps protect the panels from damage. but if you are in an area where you are likely to experience roof damage due to falling / flying objects besides the obvious of making sure your solar installation is insured at replacement value, you should ask your solar consultant / advisor regarding susceptibility to such damage to make sure that you don't lose your power when you might need it most (i.e., such storms usually cause major blackouts that sometimes can take quite a while to fully restore power.

3.3.2 **Polycrystalline solar panel**: A major physical difference between a polycrystalline and a monocrystalline solar panel is that they are rectangular with no rounded edges. Due to the imperfect crystal structure of their cell, they have a grainy bluish coating appearance. The polycrystalline panels are less efficient than the monocrystalline panels but more efficient than other types of panels. They have an efficiency rate of between 13 and 16%.

Figure 3.10 Polycrystalline solar panel

3.3.2.1 Advantages of polycrystalline solar panel

The process used to make polycrystalline silicon is simpler and costs less. The amount of waste is also less on the panel itself because of the way the silicon wafers are applied to the panel.

3.3.2.2 Disadvantages of polycrystalline solar panel

The efficiency of polycrystalline-based solar panels is a little less because of lower silicon purity. Although the difference is getting smaller all the time, you generally need to cover a slightly larger area to output the same electrical power as you would with mono panels. However, this does not mean every mono solar panel will perform better than poly panels

3.3.3 Thin film solar panel: Thin film solar cells are used mainly to power appliances with fewer power requirements due to its low efficiency. They are also less expensive and appear flexible which makes it very easy to mount on rooftops. There are three different types of Thin Film Solar cells namely:
- Amorphous Silicon (a-Si)
- cadmium telluride (CdTe)
- copper indium gallium selenide (CIS/CIGS)

These cells have an efficiency rate of between 7 and 13%.

Figure 3.11 Thin film solar panel

3.3.3.1 Advantages of a thin film solar panel

1. Versatility

Thin film can be applied to almost all types of surfaces - such as metal, plastic and even paper (in the laboratory). They've even been used as a type of roofing material. Specifically, they can actually be used instead of steel or shingles for roofing, creating an entire roof that generates power from sunlight. Unlike rigid panel types, they don't stand out, blending in better with the roof itself.

5. Flexibility

While crystal silicon solar panels are rigid and therefore fragile, "thin film" materials can be deposited on flexible substrate materials. However, while it is true that thin film solar cells are flexible, their flexibility is a feature of how they're constructed and how you can install them, but not how they're going to end up being used. Like other solar panels, they typically still get installed flat and in a frame at an optimal angle facing the sun. They can conform a little bit to a curved roof surface, but they're typically installed pretty flat; and by the time they're installed, they become inflexible.

6. Good performance in niche markets:

Thin films have long held a niche position in low power (<50W) and consumer electronics applications (e.g., calculators), and may offer particular design options for building integrated applications.

1.3.3.2 Disadvantages of a thin film solar panel

1. Efficiency

There's a reason why thin film solar panels haven't replaced older types yet. They're just not as efficient. Additionally, some thin film materials have shown degradation of performance over time and stabilized efficiencies can be 15-35% lower than initial values.

2. Longevity

You should be aware that there's a possibility that they'll hold up just as long as mono or polycrystalline panel, but this technology is new and still hasn't been well tested. There are concerns that there may be a more rapid decrease in electrical production than other types of panel technologies.

3. **Solar tracking:** Almost all of the above photovoltaic module systems have been non-solar tracking, because the output of modules has been too low to offset tracker capital and operating costs. But relatively inexpensive single-axis tracking systems can add 25% output per installed watt. This is climate-dependent. Tracking also produces a smoother output plateau around midday, allowing afternoon peaks to be met.

4. **Higher total costs:** With about a 7 to 10 percent conversion rate for energy drawn from the sun, they can only draw about half the wattage from sunlight that mono and polycrystalline panels do, which requires twice as much installation space for the same amount of power. Although panel costs (which account for around 50% of the total installed price) have been declining as a result of more efficient manufacturing and economies of scale installation costs have remained about the same. Consequently if you need to install twice as many panels to get the same results – the overall cost advantages of lower panel prices disappear quickly.

3.4. Solar panel wiring

Photovoltaic system are wired together to obtain a desired voltages, current and power rating using different wiring configurations. These configurations include;

1. Series circuit connection.
2. Parallel circuit connection.
3. Series and parallel circuit connection.

3.4.1. Series circuits connection

Two or more solar panels can be connected in series this is done by *"connecting the negative (–VE) terminal of solar panel A to the positive (+VE) terminal of solar panel B and also the negative (–VE) terminal of solar panel B to the positive (+VE) terminal of solar panel C to have a single output of a positive (+VE) and negative (-VE) terminals."*
In series connection, the voltage in the solar panel increase, the power output also increase but only the current remains the same.

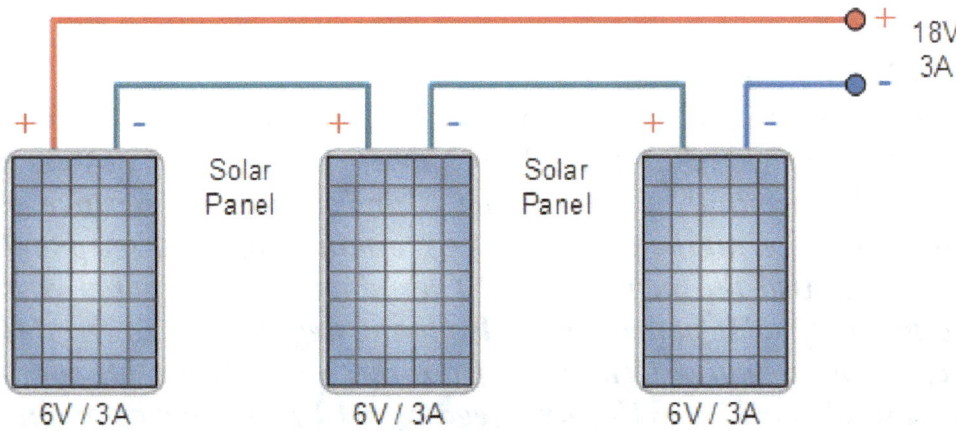

Figure 3.12 Series circuits connection

From the diagram above, three solar panels are connected in series.
Let's consider the voltages in the solar panels to be 6VDC and the current to be 3A.
When they are connected in series the voltage increases.
i.e. (6 + 6 + 6) VDC = 18VDC but, the current remains the same as 3A.
To determine the total power rating of the three panels this become:
Power (W) = Current (I) x Voltage (V)
P = 3A x 18VDC
P = 54W
This configurations can also be used on numbers of solar panels to get a desired voltages. But the current remains the same.

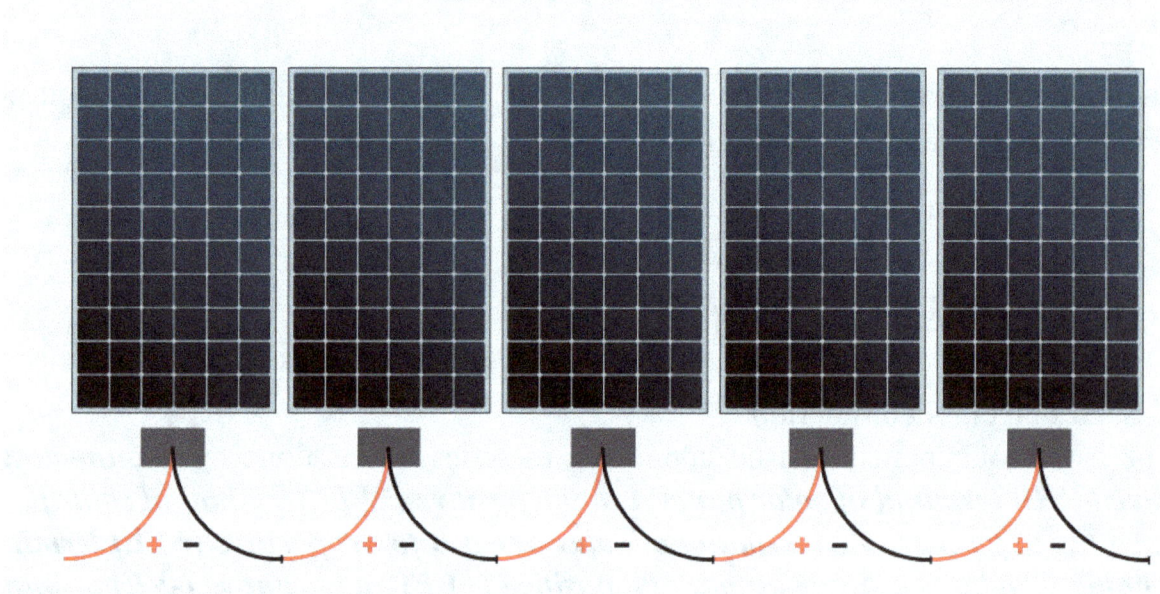

Figure 3.13 Series circuits connection

3.4.2 Parallel Circuits Connection

In the parallel connection the case is different, to connect two or more solar panels in parallel this can be achieved by *"connecting the positive (+VE) terminal of solar panel A to the positive (+VE) terminal of solar panel B and also to the positive (+VE) terminals of solar panel C i.e. all the positive (+VE) terminals of the three panels are all connected together. Also the negative (-VE) terminal of solar panel A to the negative (-VE) terminal of solar panel B and also to the negative (-VE) terminals of solar panel C i.e. all the negative terminals of the three panels are all connected together. To have a single positive (+VE) and negative (-VE) terminal as output."*

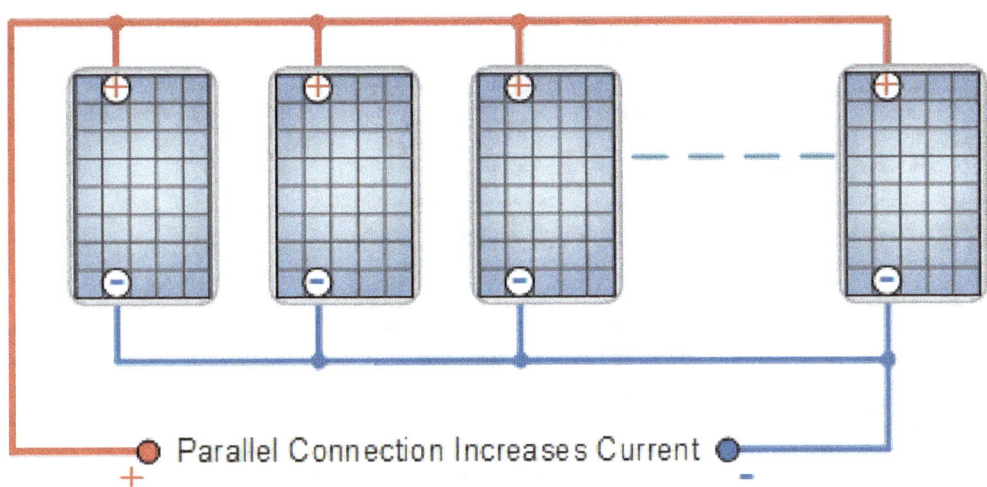

Figure 3.14 Parallel circuits connection

From the diagram above, all the positive terminals are all connected together and also all the negative terminals are all connected together.
Let's consider the voltage in each panels to be 12VDC, the current and to be 3A. The voltage remains the same i.e. unchanged while the current increases i.e. 3A + 3A + 3A + 3A = 12A and also the power rating also increases.
I.e. Power (P) = Current (A) x Voltage (V)
P = 12A X 12VDC
P = 144W

NOTE: This can also be done on numbers of solar panels to get a desired current and power output. But the voltages remains the same.

3.4.3. Series and parallel circuit connection
This is a solar panel configuration which consist of both the series and parallel connection, this configuration is done when an engineer need to meet a particular voltage and current. For example: an engineer is working on a system that requires a minimum voltage of 38VDC, Current of 32A, and a power output of 1,216W.
But given a solar panel which has a voltage 19VDC, Current 8A, and a power 152W. To achieve this aim, the engineer or installer must go with the series and parallel configuration

of solar panel.

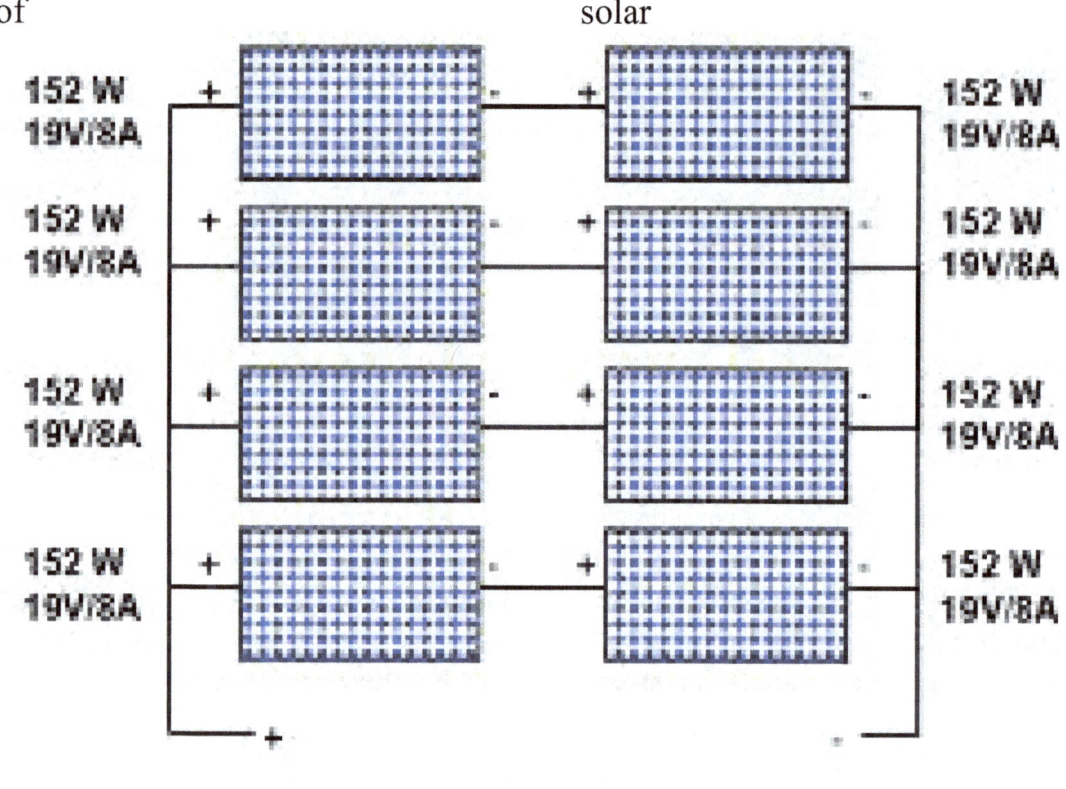

Figure 3.15 Series and parallel circuits' connection

3.5 Parameters of a solar panel

Parameters of a solar panel include the following;
1. Nominal efficiency (η):
A solar cell efficiency is the maximum output power (P_M) divided by the input power (P_{IN}). It is measured in percentage (%), which indicates that this percentage of input sunlight power is converted to electrical power. The input power is power density. Therefore, to calculate efficiency multiply P_{IN} at STC by area. The efficiency can be calculated as follows;

$$Normal\ efficiency\ (\eta) \frac{maximum\ power\ output}{input\ power}$$

$$Normal\ efficiency\ (\eta) \frac{Pm}{Pin}$$

2. **Maximum power point (Pm):**
   Maximum power point represents the maximum power that a solar cell can produce at the STC (i.e. solar radiance of 1000 W/m² and cell operating temperature of 25°C). It is measured in W_{Peak} or simply W_P. Other than STC the solar cell has P_M at different values of radiance and the cell operating temperature.
   $P_M = I_M \times V_M$

3. **Maximum power voltage (Vmp):**
   It represents the voltage that the solar cell will produce when operating at the maximum Power Point. It is denoted by V_M.

4. **Maximum power current (Imp):**
   It represents the current which the solar cell will produce when operating at the maximum PowerPoint. It is denoted by I_M.

5. **Open circuit voltage (Voc):**
   It represents the voltage that the solar cell will produce when operating at the maximum PowerPoint. It is denoted by V_M.

6. **Short circuit current (Isc):**
   Short circuit current is the maximum current produced by the solar cell, it is measured in ampere (A) or milli-ampere (mA).

7. **Peak power (Pmaxs):**
   The maximum electric or nominal power of your PV system can be defined as its 'Peak Power' (in Watt Peak).

8. **Power tolerance (Tol):** Power Tolerance is the actual range a module can deviate from its specified STC Max Power. STC stands for Standard Test Conditions, and refers to the lab conditions panels are tested under (1000W per square meter of sunlight).

   For example, if a 220W panel has a Power Tolerance of -10% / +10%, the panel could have an actual Max Power of anywhere between 198 and 242W.

   As the technology behind solar panels continues to advance, it's becoming more common for panels to have a 0% or 0W negative power tolerance. This means the solar panel will always have a rated STC Max Power equal or greater than what's been specified.

9. **Application class:** The electrical shock hazard tests will differ depending on the application class of the solar panel being tested. To correctly forecast the test program for the module, the application class must be determined:

- **Application class A**: for modules operating at greater than 50 VDC or 240 W and in an area where the module can be contacted by the public. This PV panel class has the most stringent testing program.
- **Application class B:** for modules installed in locations not accessible by the public. It has a slightly less stringent test program than the Class A modules, but the differences in testing programs are not significant.
- **Application class C:** for modules operating at less than 50 VDC and 240 W. These less powerful PV panels have a significantly reduced test program.

10. **Maximum system fuse ratting:** The fuse ratting for a solar panel is usually 15A at maximum, but if the engineer is not sure but the DC cable has been sized correctly then the fuse rating is the maximum current x 10%. Also, in a case where the engineer is not sure of which wire to use, a 12 gauge wire is best to use.

3.6 Load resistance

The load or battery determines the voltage at which the solar panel will operate. For example, let's consider a 12VDC battery, the battery voltage is usually between the ranges of 11.5VDC to 13VDC. For the batteries to be charged, the solar panel must operate at a higher voltage than the battery voltage. I.e. if the battery operates at 12VDC the solar panel must operate at a minimum voltage of 17VDC and above, or else the battery will not charge.

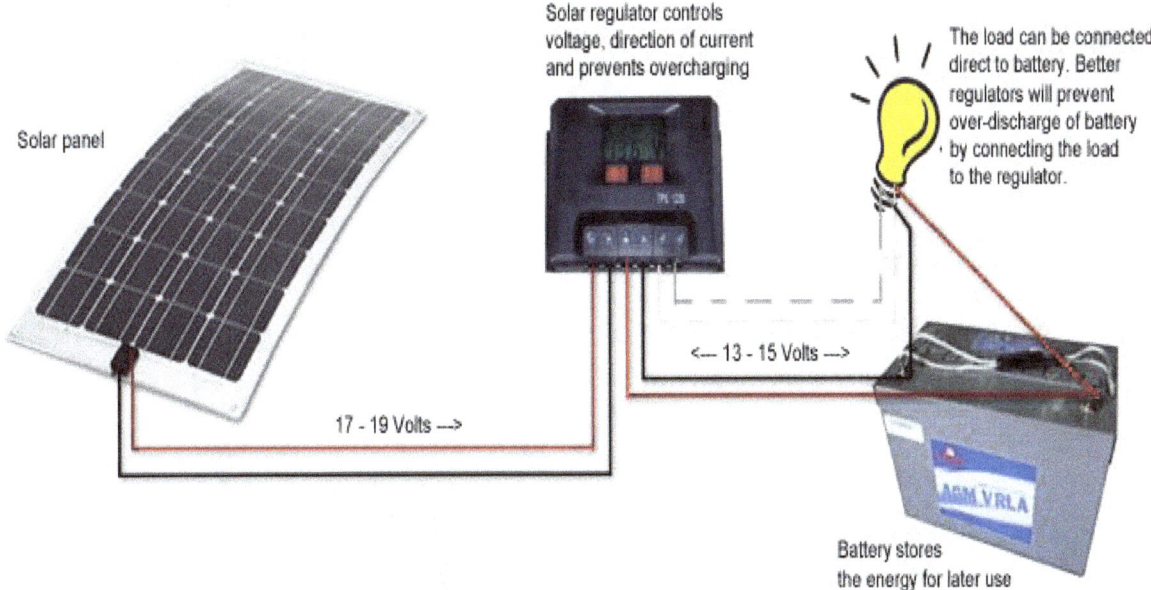

Figure 3.16 Solar panel voltage higher than battery voltage

When the solar panel voltage is higher than the battery voltage. The battery will charge faster, as shown in the image above.

Also, when the battery voltage is higher than the solar panel voltage the battery will never charge, as shown in the image below.

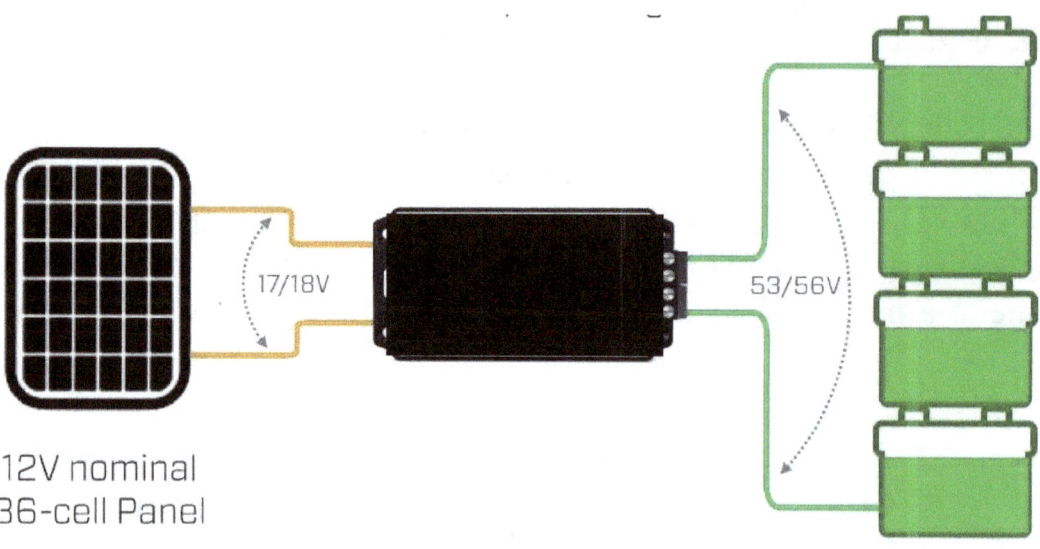

Figure 3.17 Battery voltage higher than solar panel voltage

Also when, the battery voltage and solar panel voltage are the same, the battery will also not charge or might charge at a very slow rate.

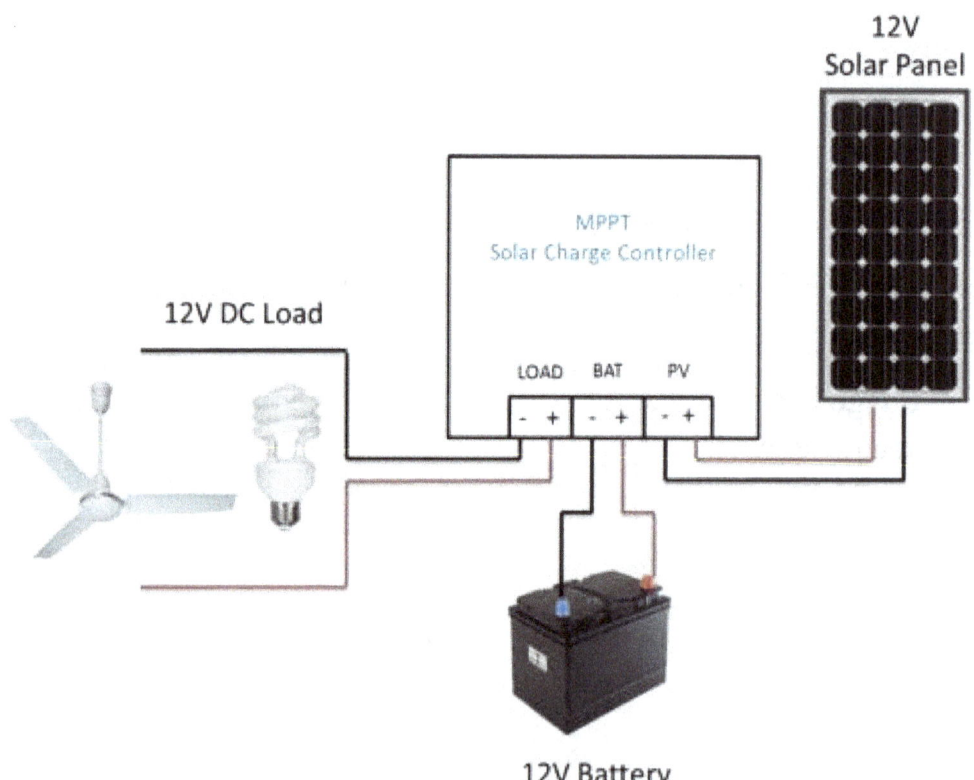

Figure 3.18 Solar panel and battery having same voltage

3.7. Effect of solar luminance or intensity

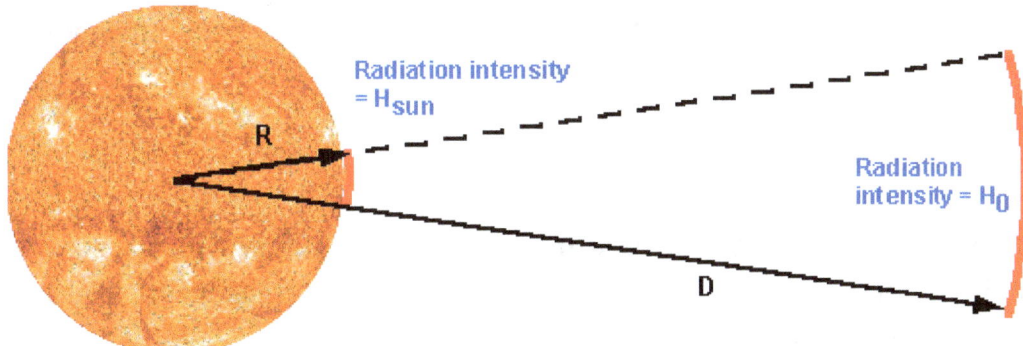

Figure 3.19 Solar luminance or intensity

Solar cells experience daily variations in light (sun) intensity, with the incident power from the sun varying between 0 and 1 kW/m^2. At low light levels, the effect of the shunt resistance becomes increasingly important. As the light intensity decreases, the bias point and current through the solar cell also decreases, and the equivalent resistance of the solar cell may begin to approach the shunt resistance. When these two resistances are similar, the fraction of the total current flowing through the shunt resistance increases, thereby increasing the fractional power loss due to shunt resistance. Consequently, under cloudy conditions, a solar cell with a high shunt resistance retains a greater fraction of its original efficiency than a solar cell with a low shunt resistance.

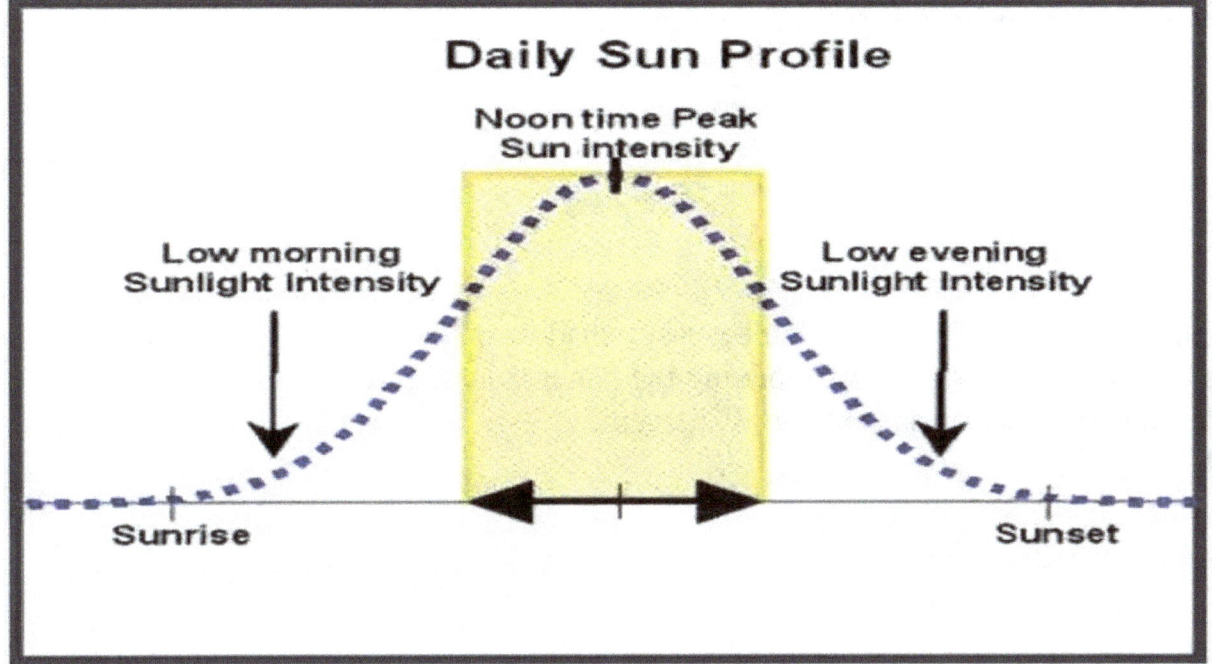

Figure 3.20 Daily sun profile

The diagram above shows that the open circuit voltage, short-circuit current, and maximum output power of solar cells increase with the increase of light (sun) intensity. Therefore, it can be known that the greater the light intensity, the better the power generation performance of the solar cell vise-vasa.

3.8. Cell temperature

The higher the cell temperature *(cell temperature not ambient air temperature)* of a solar panel the lower the output voltage, once the cell temperature increases above 25 degree Celsius there will be a voltage drop.

Figure 3.21 Cell temperature

3.9. Shading

Shading of solar panel is when an object cast a shadow on the solar panel or shading is when there is an obstructions blocking the panel from receiving direct sunlight, which reduces its performance up to 75 percent.

Figure 3.22 Shaded solar panel

Solar PV panels are very sensitive to solar shadings. Total or partial shading conditions have a significant impact rate on the capability of delivering energy and may result in lower output and power losses. Cells in a solar panel are usually connected in series to get a higher voltage and therefore an appropriate production of electricity. But when shading occurs, this structure presents some limitations.

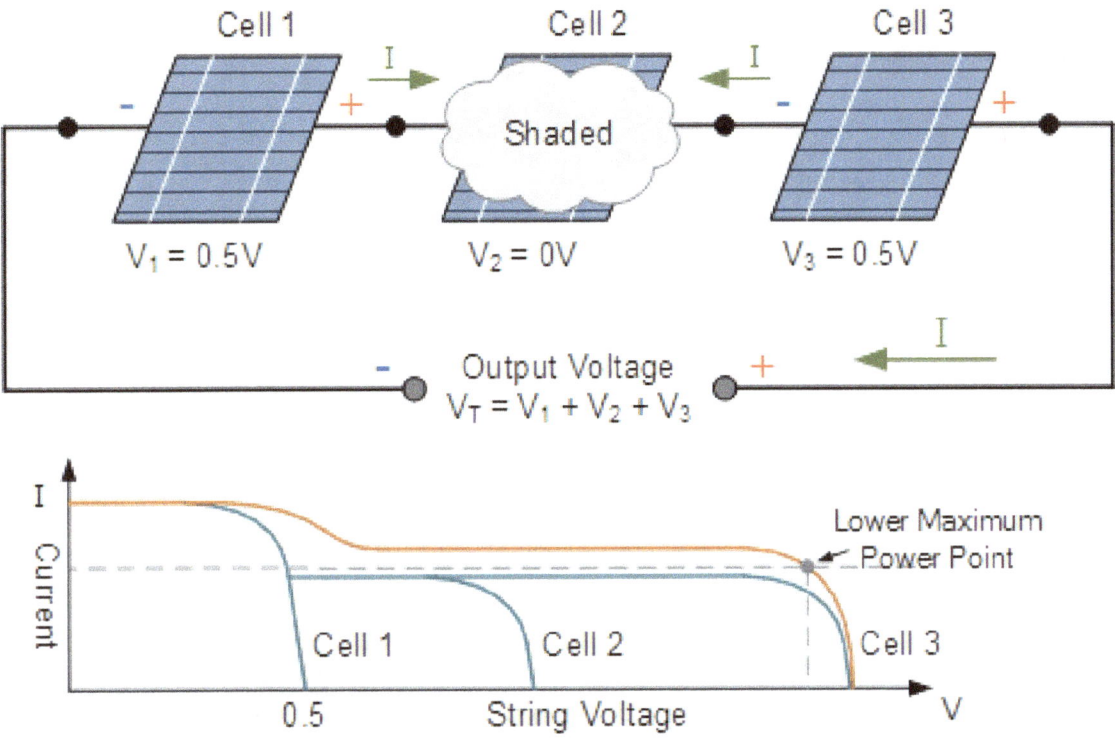

Figure 3.23 Current flow in a shaded solar cell

In fact, when a single solar cell is shaded, the current of all the units in the string is determined by the unit that produces the least current. When a cell is shaded, the whole series is virtually shaded too.

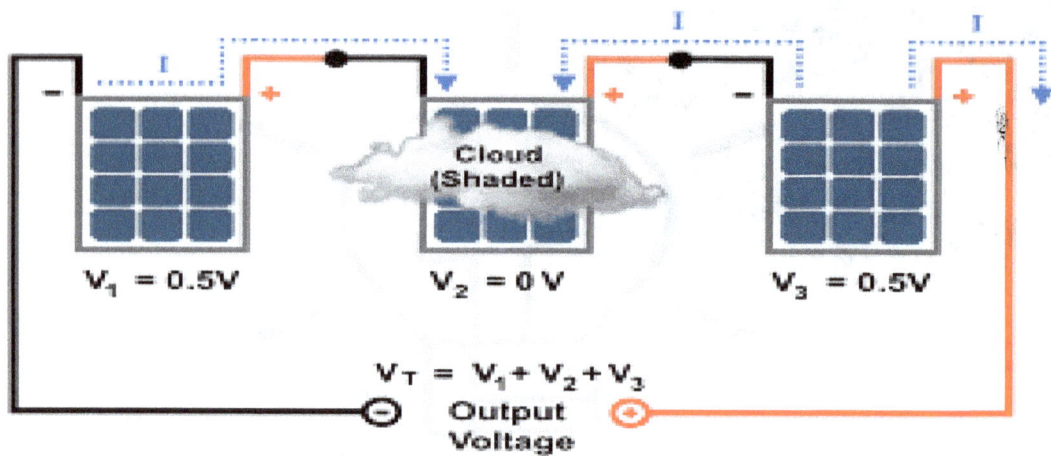

Figure 3.24 Current flow in a shaded solar cell

NOTE: To prevent the loss of energy due to shading, the installation usually includes bypass diodes. Bypass diodes are wired in parallel to the solar cells. When a solar cell is shaded, the bypass diode provides a current path which allows the string of connected solar cells generating energy at a reduced voltage.

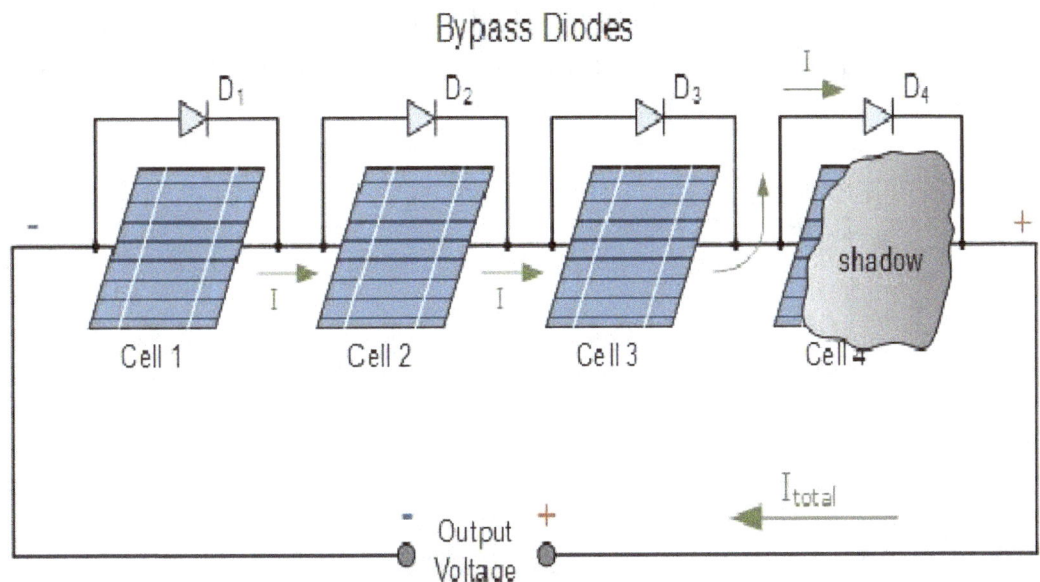

Figure 3.25 Bypass diode

3.10. Diodes

A diode is a semiconductor device that allows electric current to pass in only one direction.

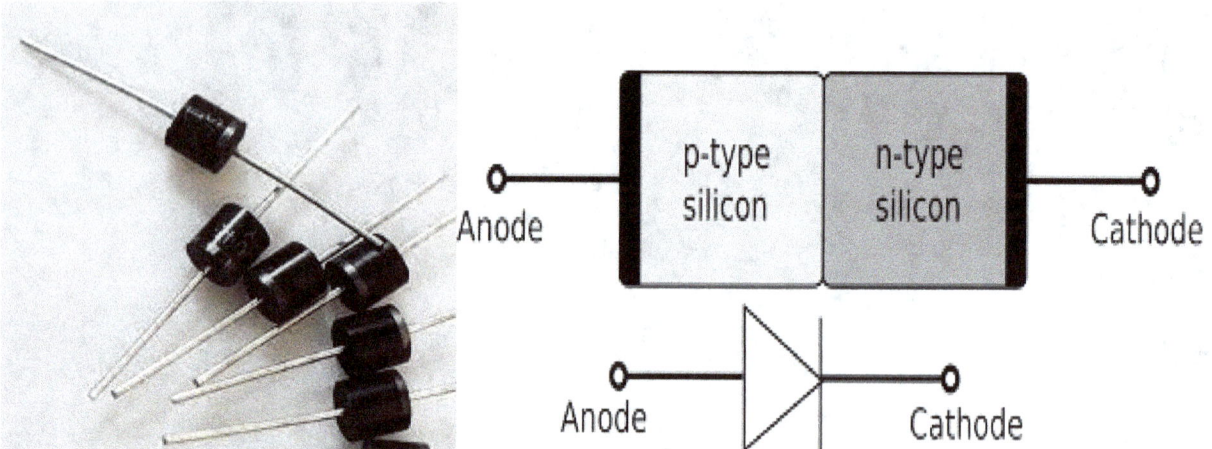

Figure 3.26 Diode

In a solar panel, diodes may be used for several functions as follows:

3.11. Blocking diodes

This type of a diode is placed in the positive line between the modules and the battery to prevent reverse current flow from the batteries to the array at night or during cloudy weather. Some controllers already contain a diode or perform this function by opening the circuit.

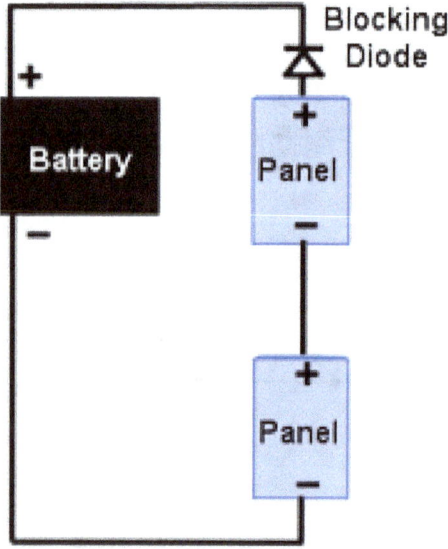

Figure 3.27 Blocking diode

NOTE: Do not remove the blocking diode between the solar panel; it will damage your panel by reverse current flow.

3.12. Bypass diodes

This type of a diode is wired within a solar panel to divert current around a few cells in the event of shading.

Figure 3.28 Bypass diode in a junction box

CHAPTER FOUR

4.1. Working principle of a lead acid battery

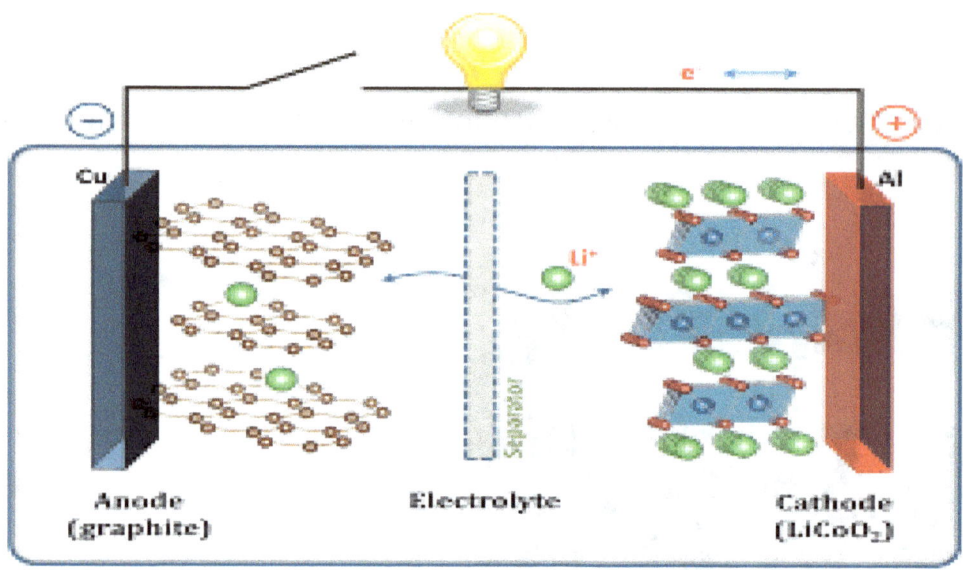

Figure 4.1 Working principle of a lead acid battery

In a lead acid battery, as sulphuric acid is commonly used as an electrolyte in the battery, when it gets dissolved, the molecules in it are dispersed as SO_4^- (negative ions) and 2H+ (positive ions) and these will have free movement. When these electrodes are dipped in the solutions and provide a DC supply, then the positive ions will have a movement and move towards the direction of the negative edge of the battery. In the same way, the negative ions will have a movement and move towards the direction of the positive edge of the battery. Every hydrogen and sulfate ions collect one and two electron and negative ions from the cathode and anode and they have a reaction with water. This forms hydrogen and sulphuric acid. Whereas the developed from the above reactions react with lead oxide and forms what is called lead peroxide. This means at the time of the charging process; the lead cathode element stays as lead itself whereas the lead anode is formed as lead peroxide which is dark brown in color.

When there is no DC supply and then at the time when a voltmeter is connected in between the electrodes, it displays the potential difference between electrodes. When there is a connection of wire between the electrodes, there will be a passage of current from the negative to the positive plate through an external circuit which signifies that the cell holds the ability to provide an electric form of energy. This phenomenon is known as the working principle of a lead acid battery.

4.2. Types of batteries and their operation

There are different types of batteries used for solar and inverter system which include;

4.2.1 Dry cell battery: Dry cell batteries refer to those batteries that use an extremely low moisture electrolyte. Furthermore they contrast with lead-acid batteries that use a liquid electrolyte.

Figure 4.2 Dry cell battery

4.2.1.1 Advantages of lead acid or dry cell batteries
1. It is not expensive
2. It is simple to manufacture
3. High power output capacity
4. Easily rechargeable.
5. High charge and discharge current.
6. Minimal corrosion
7. It can be installed in non-ventilated area

4.2.1.2 Disadvantages of lead acid batteries
1. Lower energy density
2. It is very heavy
3. Short life span
4. Not suitable for heavy duty application.

4.2.2 Tubular battery: It is also called lead acid battery. The lead-acid batteries consist of two electrodes dipped in the electrolyte solution. The electrodes are lead and lead dioxide, hence the name lead-acid batteries. These two electrodes immersed in sulphuric acid causes the chemical reactions that generate DC current.

They are the most efficient inverter batteries and are quite popular as well. The pencil-type armored tubular plates are designed for uninterrupted power delivery in case of long power cuts.

Figure 4.3 Tubular battery

The tubular inverter battery for home comes in abrasion-resistance, leak-proof tower type, which ensures lower maintenance and better safety. These batteries are safe for the environment, and their efficient working makes them an ideal buy for the homes.

4.2.2.1 Advantages of tubular batteries
1. High efficiency.
2. Highly reliable.
3. Long life span.
4. Fast recharge.
5. It can be topped with distilled water to prolong life span.
6. Suitable for heavy-duty application.

4.2.2.2 Disadvantages of tubular batteries

1. Very expensive.
2. It is very heavy.
3. It must be installed in a ventilated area.

4.2.3 Lithium ion battery: A lithium-ion (Li-ion) battery is an advanced battery technology that uses lithium ions as a key component of its electrochemistry. Lithium batteries stand apart from other battery chemistries due to their high energy density and low cost per cycle. However, "lithium battery" is an ambiguous term. There are about six common chemistries or types of lithium batteries, all with their own unique advantages and disadvantages.

4.2.4 Types of lithium-ion batteries

1. **Lithium cobalt oxide:** has high specific energy compared to the other batteries, making it the preferred choice for laptops and mobile phones. It also has a low cost and a moderate performance. However, it is highly unfavorable in all the other aspects when compared to the other lithium-ion batteries. It has low specific power, low safety, and a low lifespan.

Figure 4.4 Lithium cobalt oxide

2. **Lithium manganese oxide**: has moderate specific power, moderate specific energy, and a moderate level of safety when compared to the other types of lithium-ion batteries. It has the added advantage of a low cost. The downsides are its low performance and low lifespan. It is usually used in medical devices and power tools.

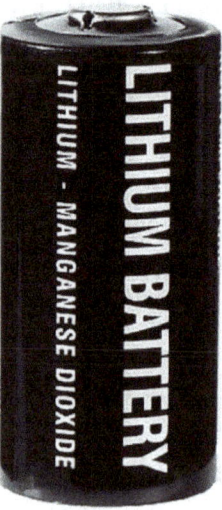

Figure 4.5 Lithium manganese oxide

3. **Lithium nickel manganese cobalt oxide:** has two major advantages as compared to the other batteries. The first one is its high specific energy, which makes it desirable in electric powertrains, electric vehicles, and electric bikes. The other is its low cost. It is moderate in terms of specific power, safety, lifespan, and performance when compared to the other lithium-ion batteries. It can be optimized to either have high specific power or high specific energy.

Figure 4.6 lithium nickel manganese cobalt oxide

4. **Lithium iron phosphate (lifepo4):** Has one major disadvantage when compared to other types of lithium-ion batteries, and that is its low specific energy. Other than that, it has moderate to high ratings in all the other characteristics. It has high specific power, offers a high level of safety, has a high lifespan, and comes at a low cost. The performance of this battery is also moderate. It is often employed as storage for solar and inverter system, electric motorcycles, and other applications that require a long lifespan and a high level of safety.

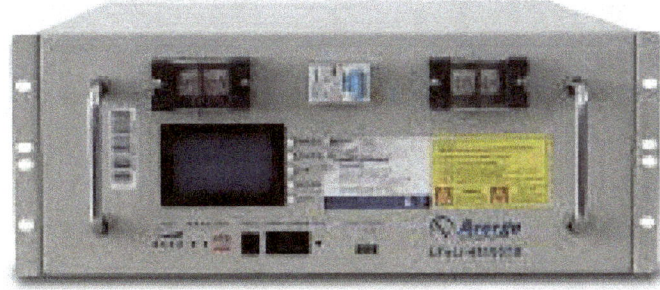

Figure 4.7 Lithium iron phosphate (lifepo4)

5. **Lithium nickel cobalt aluminum oxide**: offers one strong advantage compared to the five other batteries high specific energy. It is pretty moderate in the rest of the characteristics like performance, cost, specific power, and lifespan. The only downside to this battery type is its low level of safety. Its high specific energy and moderate lifespan make it a good candidate for electric powertrains.

Figure 4.8 Lithium nickel cobalt aluminum oxide

6. **Lithium titanate:** offers high safety, high performance, and a high lifespan which are very important features every battery should have. Its specific energy is low compared to the five other lithium-ion batteries, but it compensates for this with moderate specific power. The only major disadvantage of lithium titanate as compared to the other lithium-ion batteries is its extremely high cost.

Figure 4.9 Lithium titanate

Another important feature of this battery worthy of mention is its remarkably fast recharge time. It can be used for storing solar energy and creating smart grids.

Comparison between the six types of lithium ion batteries

Lithium-ion battery types	SP	SE	SF	LS	CS	PF
Lithium cobalt oxide	L	H	L	L	L	M
Lithium manganese oxide	M	M	M	L	L	L
Lithium nickel manganese cobalt oxide	M	H	M	M	L	M
Lithium iron phosphate	H	L	H	H	L	M

Lithium nickel cobalt aluminum oxide	M	H	L	M	M	M
Lithium titanate	M	L	H	H	H	H

- **SP** stands for specific power
- **SE** stands for specific energy
- **SF** stands for safety
- **LS** stands for lifespan
- **CS** stands for cost
- **PF** stands for performance
- **L** stands for low
- **M** stands for moderate
- **H** stands for high

Table 4.1 *Comparison between the six types of lithium ion batteries*

4.2.2.1. Advantages of lithium ion battery
1. Longest life span
2. Last longer than any other battery
3. No maintenance
4. It can be installed in a non-ventilated area
5. The monomer capacity is 5AH-1000AH i.e. it has a larger capacity than ordinary batteries (lead-acid, etc.).
6. Eco friendly

4.2.2.2. Disadvantages of lithium ion battery
1. Too expensive
2. Too heavy
3. Low nominal voltages
4. It require technical experts during installation

4.3. Days of autonomy

Autonomy refers to the number of days a battery system will run a given load (appliances) without being recharged by the solar panel or from the grid.

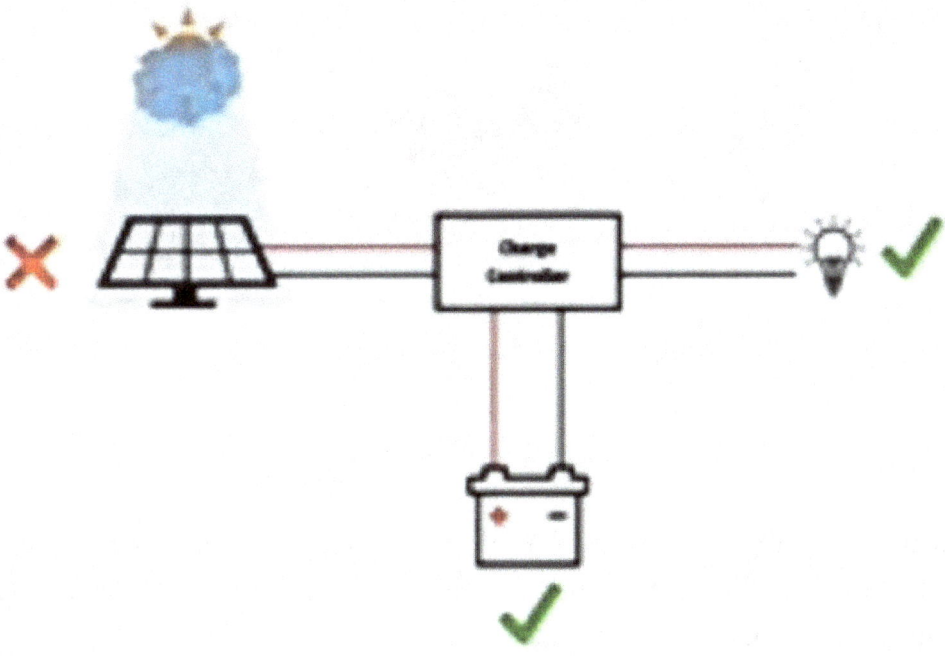

Figure 4.10 Days of autonomy

To achieve this you must consider the following

- Geographical location of the site.
- Energy consumption or load audit.
- Types of loads.

By knowing all these will enable the engineer to determine the days of autonomy needed.

4.3.1 Geographical location of the site.

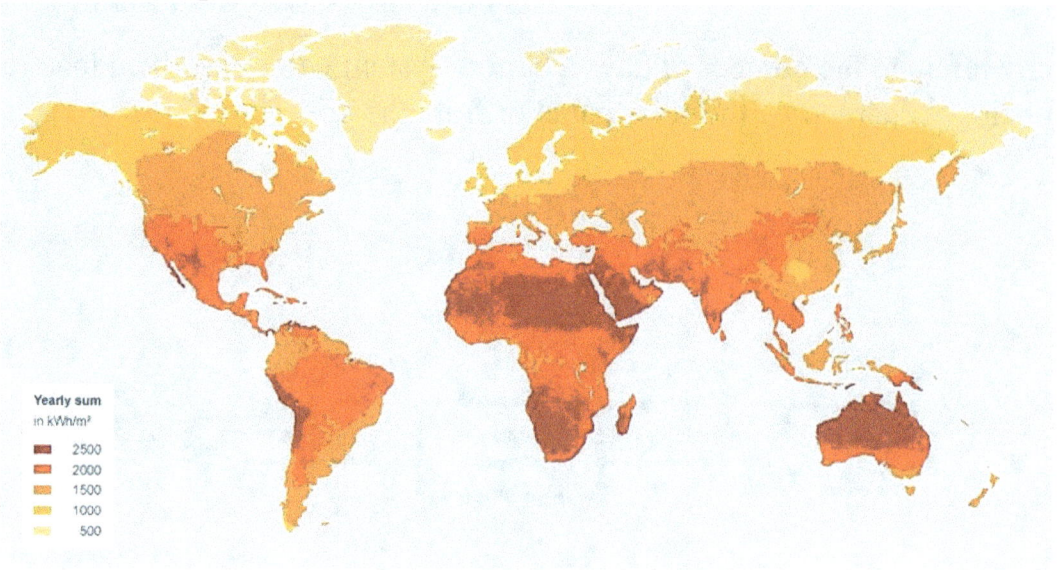

Figure 4.11 World map

Generally, the weather condition of a geographical area can determine the number of "no sun" days which is an important variable in determining the days of autonomy. Using Nigeria as a case study, in Nigeria there are two seasons such as the **rainy season (April to September) and dry season (December to March)**. During the dry season there are excessive sun shine in most parts of the country with little or no cloudy sky. But on the other hand the raining season is always cloudy with little or no sun shine which will reduce the hours of sunshine needed to charge the battery by the solar panels (photovoltaic system). So during this season the days of autonomy must be considered

4.3.2 Energy consumption or load audit:

The main aim of a Solar Energy Audit is to know the energy demand of the client. An energy audit is not a design or works specification. However, data from an energy audit is required for the designing of a solar system solution. Without an accurate energy audit, it's seemingly impossible to design a sustainable solar system.

Sr. No.	Equipment Name	Number of Equipments	Wattage	Total Wattage	Working Hours	Watt hours
1.	Tube light	3	40	120	10	1200
2.	Fan	3	80	240	10	2400
3.	Television	1	150	150	5	750
4.	Fridge	1	350	350	12	4200
5.	Air Conditioner	1	1500	1500	5	7500
	Total		2120	2360		16,050

Table 4.2 Load audit

Total energy consumption = Quantity of electrical appliance x Power rating of appliance x daily running hours.

Energy Audit of a house, office is a careful examination of the electrical loads. This is crucial for stand-alone system which are commonly seen in Nigeria. Load audit helps prevent over sizing of the system, safe guarding your electrical appliances from damaging from inappropriate power surge and it generally saves your cost.

What to consider during Solar Energy Audit
1. Total energy consumption of the house or office.
2. List of electrical appliances to be powered by the solar system.
3. Power rating of the Electrical appliances in Watt.

4.3.3 Type of load

Figure 4.11 Load type

The term "load" refers to the power consumption of the devices that are being used in the system. Understanding your loads is critical to maintaining a well-functioning solar system.

The problem with many clients in Nigeria is that they don't understand their own loads. For example, an off-grid solar system may power a fridge, air condition, or both at the same time. The power being consumed by these devices are the system's "loads". The fridge may consume 350 Watts of power, while the air condition may consume 800 Watts of power, resulting in a total continuous load of 1,150 Watts, if using both simultaneously, which of course is very straight forward. But, some more complicated sites may have multiple combinations of the above example. If necessary, calculate each load individually and then combine them to calculate the total load for the site. If necessary, a spreadsheet can help to keep track of the loads and the calculations.

4.4. Battery capacity

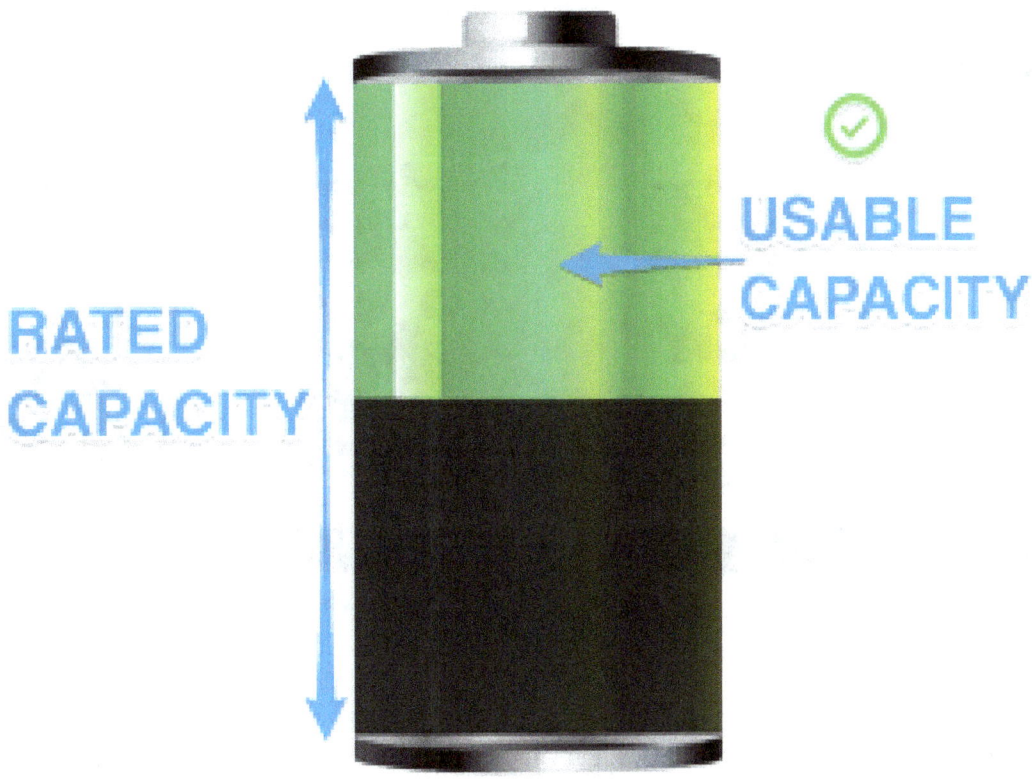

Figure 4.12 Battery capacity

The energy stored in a battery, called the battery capacity, is measured in either watt-hours (Wh), kilowatt-hours (kWh), or ampere-hours (Ah) as the case may be. The most common measure of battery capacity is in Ah, which is the number of hours for which a battery can provide a current equal to the discharge rate at the nominal voltage of the battery.

For example; if a 100 AH battery will deliver one amp for 100 hours or roughly two amps for 50 hours before the battery is considered fully discharged. If more storage capacity is required to meet a specific application requirement, such batteries can be connected in parallel. Two 100 amp-hour 12-volt batteries wired in parallel provide 200 amp-hours at 12 volts. Higher voltages are obtained through series wiring. Two 100 amp-hour 12 volt batteries wired in series provide 100 amp-hours at 24 volts.

4.5 The rate and depth of discharge

4.5.1 Rate of discharge:

Figure 4.13 Rate of discharge

The discharge rate is given by the battery capacity (Ah) divided by the number of hours it takes to charge/discharge the battery. For example, If a battery capacity of 500 Ah that is theoretically discharged to its cut-off voltage in 20 hours will have a discharge rate of 500 Ah/20 h = 25 A. Furthermore, if the battery is a 12VDC battery, then the power being delivered to the load is 25A x 12 VDC = 300W.

Note: A battery can only "theoretically" discharge to its maximum level as most practical batteries cannot be fully discharged without either damaging the battery or reducing its lifetime.

4.5.2. Depth of discharge (DOD):

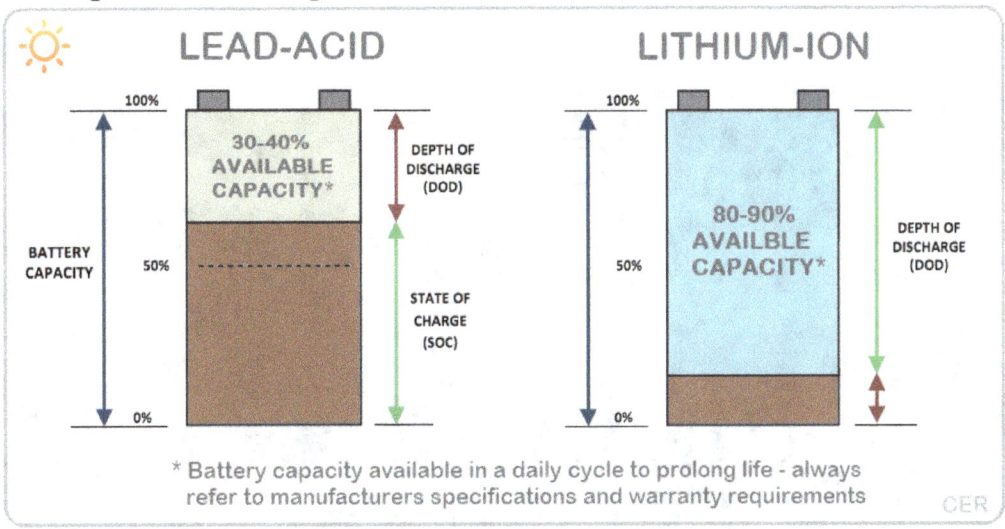

Figure 4.14 Depth of discharge

"When the nickel cadmium of a lithium-ion battery is fully discharged it may reverse polarity, potentially harming the load".

The depth of discharge refers to how much capacity will be withdrawn from a battery. Most PV systems are designed for regular discharges of 40 percent to 80 percent. Battery life is directly related to how deep the battery is cycled. For example, if a battery is discharged 40 percent every day, it will last about twice as long as if it is discharged 80 percent. Lead-acid batteries should never be completely discharged; even though some deep cycle batteries can survive this condition, the voltage will continually decrease. Lithium-ion batteries, on the other hand, can be totally discharged without harming the battery and hold their voltage.

4.6 Battery safety

Figure 4.15 Battery hazard

Batteries used for solar panel installation are one of the most dangerous types of batteries, if they are not properly handled, installed, maintained. Due to Dangerous chemicals, heavy weight, and high voltages and currents are potential hazards that can result in electric shock, burns, explosion, or corrosive damage.

The following precautions should be taken when installing, replacing, and carrying out maintenance on photovoltaic battery systems:

Do's:

1. Make sure other components of the system are switched off before refilling tubular batteries with a distilled water. I.e. ensure that the inverter, PV breaker, grid breaker and other source of energy are all switched off to avoid electrical shock.

Figure 4.16 Refiling lead acid battery

2. While connecting the inverter and charge controller terminals to the battery, ensure it is properly tightened and not loosed, to prevent partial contact that can cause damages to the battery and other components in the system.

7. Use proper tools when assembling cells.

8. Remove all jewelries or any metallic objects that can bridge the battery terminals to avoid electric shock. Tools such as combination spanner, socket spanner should be properly insulated to avoid electric shock. Because such tools can easily bridge battery terminals.

9. Make sure baking soda is available to neutralize acid spills on your skin.

10. Wear protective equipment's to protect your eye and skin from acid burn.

Figure 4.17 Safety equipment

11. Other batteries can be installed in a non-ventilated area. But tubular batteries must be installed in a ventilated area.

12. Discharge body static electricity before touching terminal posts
13. Lift batteries from the bottom or use carrying straps.
14. Prevent metals from falling on terminals of the battery, otherwise battery can catch fire.
15. Use common sense.

Don'ts:

1. Do not use non-insulated metal tools while installing batteries or carrying out maintenance.
2. Do not smoke near battery.
3. Do not lift batteries by their terminal ports or by squeezing the sides of the battery.

4.7. Battery Wiring Configuration

Batteries need to be configured to obtain the desired voltage and amp-hours.

There are three types of battery circuit configuration or connections;
1. Series circuit connection.
2. Parallel circuit connection.
3. Series and parallel circuit connection.

4.7.1 Series circuit connection:

The series circuit connection of a battery is similar to the series circuit connection of a solar panel. Which involve the connection of the positive (+VE) terminal of battery A to the negative (-VE) terminal of battery B to have a single output. In this kind of connection the voltage increase while the current remains the same.

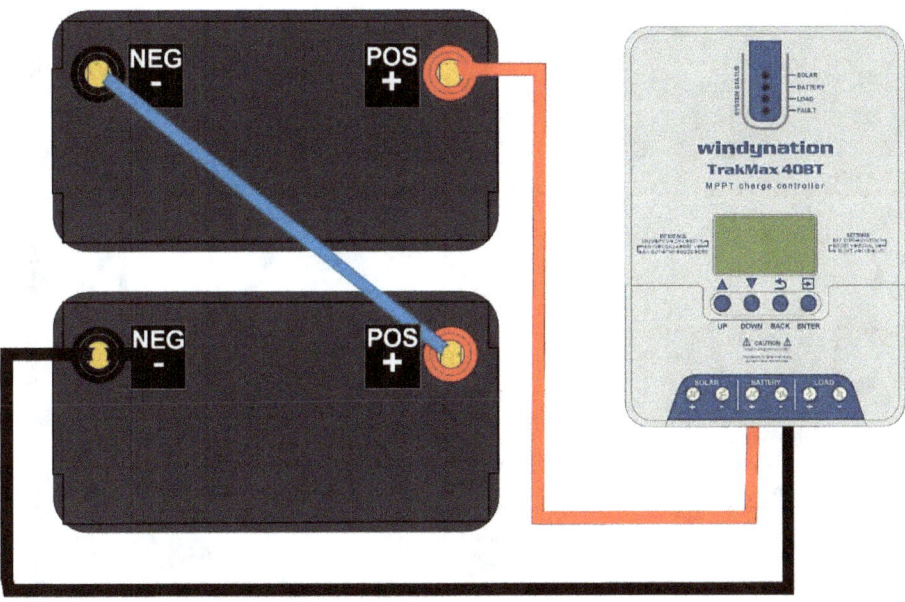

Figure 4.18 series circuit connection

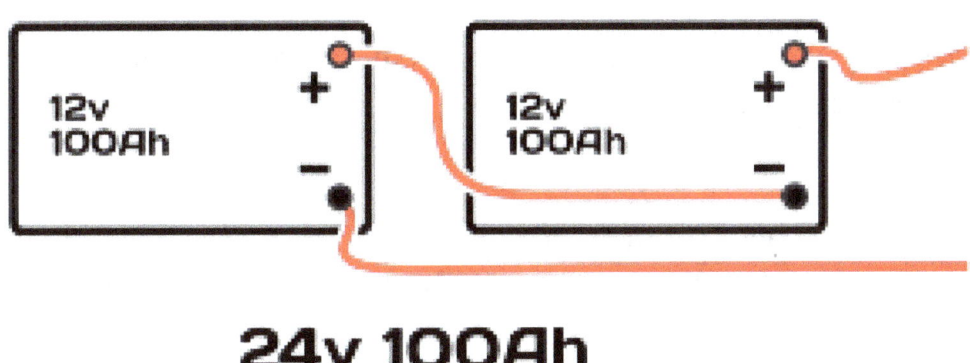

Figure 4.19 Series circuit connection

From the diagram above, two 12VDC, 100AH batteries are connected in series the positive (+VE) terminal of battery A connected to the negative (-VE) terminal of battery B to have a single output of 24VDC. But the amperage (current) remains as 100AH.

The result or output gives us 24VDC 100AH battery bank.

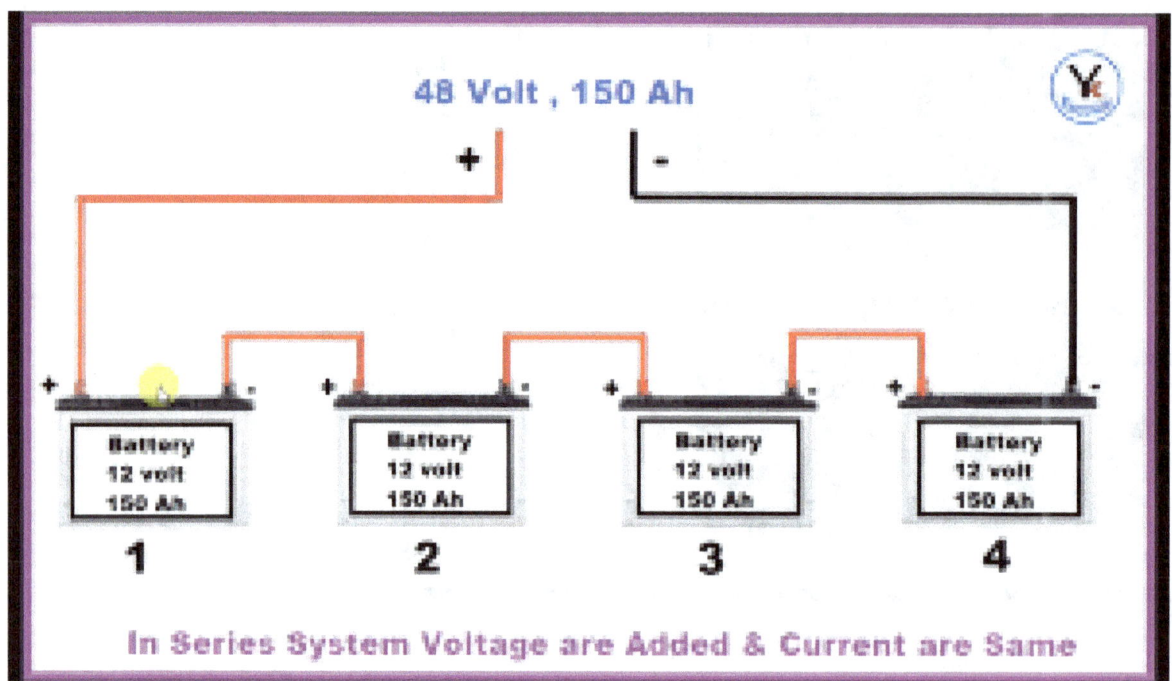

Figure 4.20 Series circuit connection

More batteries can be connected in series to obtain desired voltages, but the amperage (current) remains the same.

4.7.2 Parallel circuit connection:

In parallel connection the positive (+VE) terminal of battery is connected to the positive (+VE) terminal of battery B also the negative (-VE) terminal of battery A is connected to the negative (-VE) terminal of battery B. to have a single output.

This connection will increase the amperage (current) but the voltage remains the same.

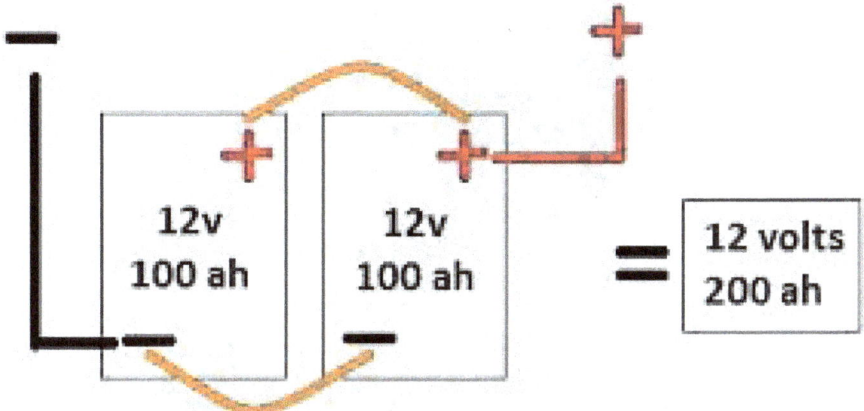

Figure 4.21 Parallel circuit connection.

From the diagram above, two 12VDC 100AH batteries are connected in parallel in such a way that the positive (+VE) terminal of battery A is connected to the positive (+VE) terminal of battery B, and the negative (-VE) terminal of battery A is connected to the negative (-VE) terminal of battery B.

The amperage (current) increases while the voltage remains the same.
The result or output gives us 12VDC 200AH battery bank.

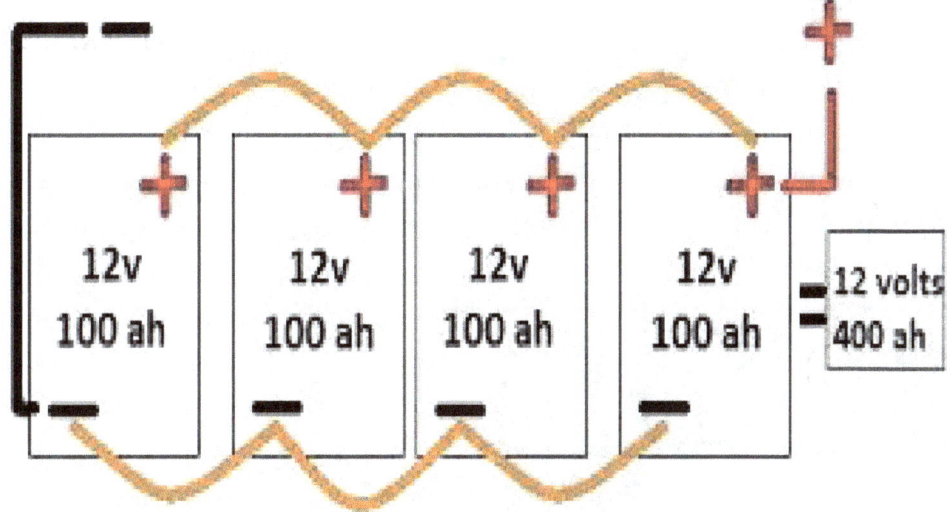

Figure 4.22 parallel circuit connection

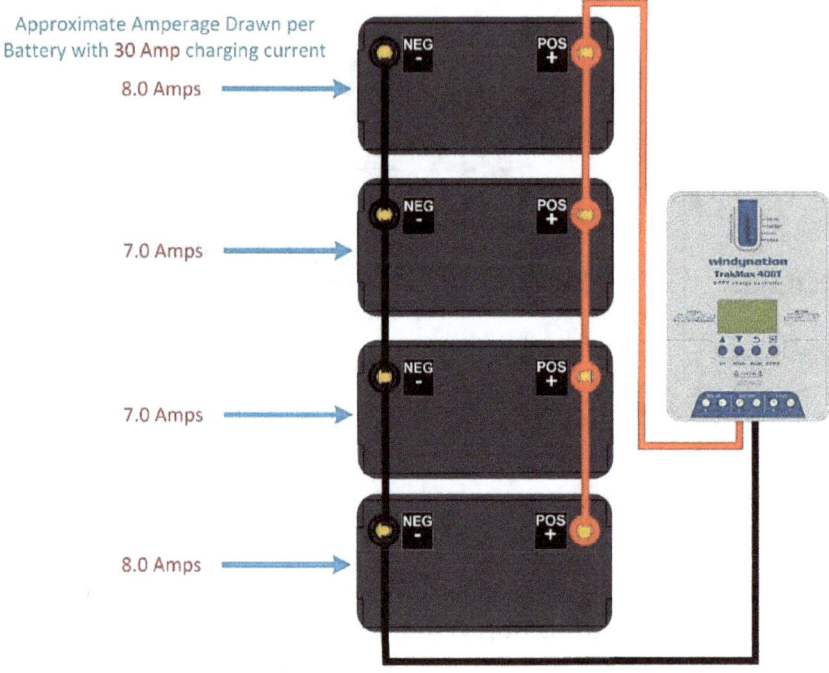

Figure 4.23 Parallel circuit connection

More batteries can also be connected in parallel to obtain a desired amperage (current) but the same voltage, as shown in figure 4.22 and figure 4.23.

4.7.3 Series and parallel circuit connection

The series and parallel circuit connection is a kind of connection which have both the series configuration and the parallel configuration together to achieve a desired voltage or current.

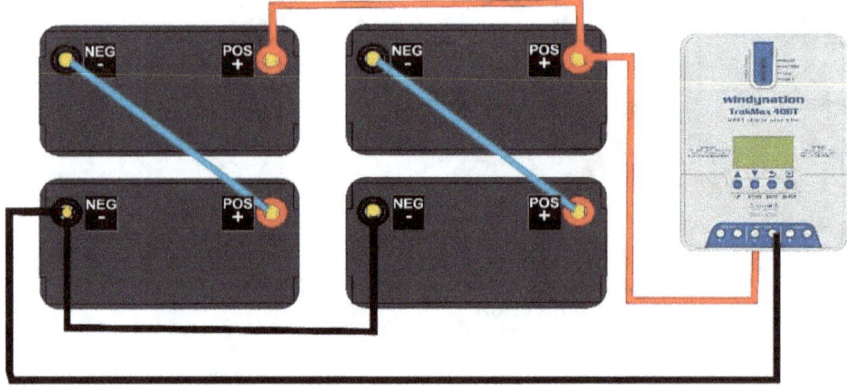

Figure 4.24 Series and parallel circuit connection

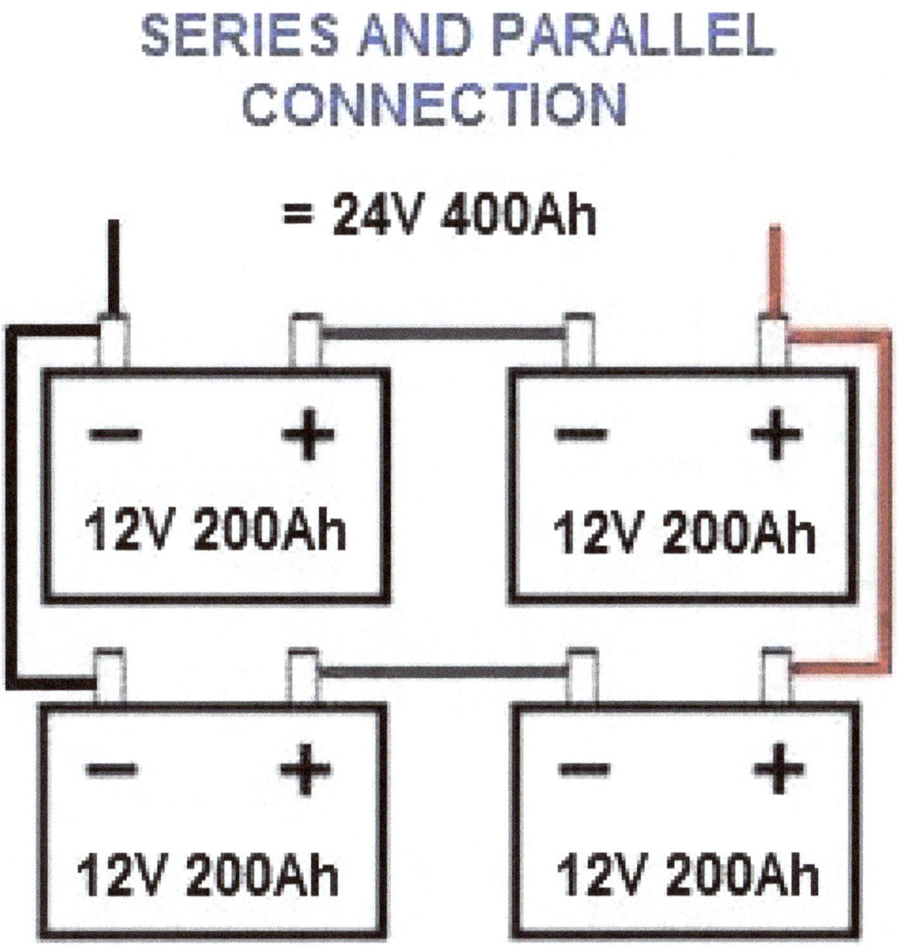

SERIES AND PARALLEL CONNECTION

= 24V 400Ah

Figure 4.25 Series and parallel circuit connection

From the diagram above, an engineer was given four 12VDC 200AH batteries to work on a system that requires 24VDC 400AH.

In other to achieve this aim, he must first connect two 12VDC 200AH batteries in series to obtain 24VDC 200AH battery bank. Also he will connect another 12VDC 200AH battery is series to obtain another 24VDC 200AH battery bank.

But, what is needed by the system is 24VDC 400AH.

By connecting both 24VDC 200AH battery banks in parallel we will have a 24VDC 400AH battery bank needed by the system.

More batteries can be connected in series and parallel to achieve a desired voltage and current needed for a given system.

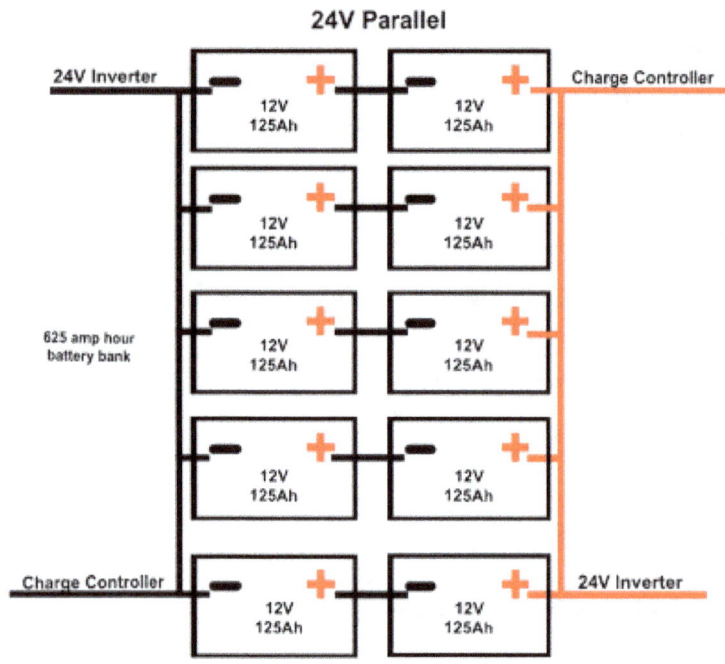

Figure 4.26 Series parallel circuit connection

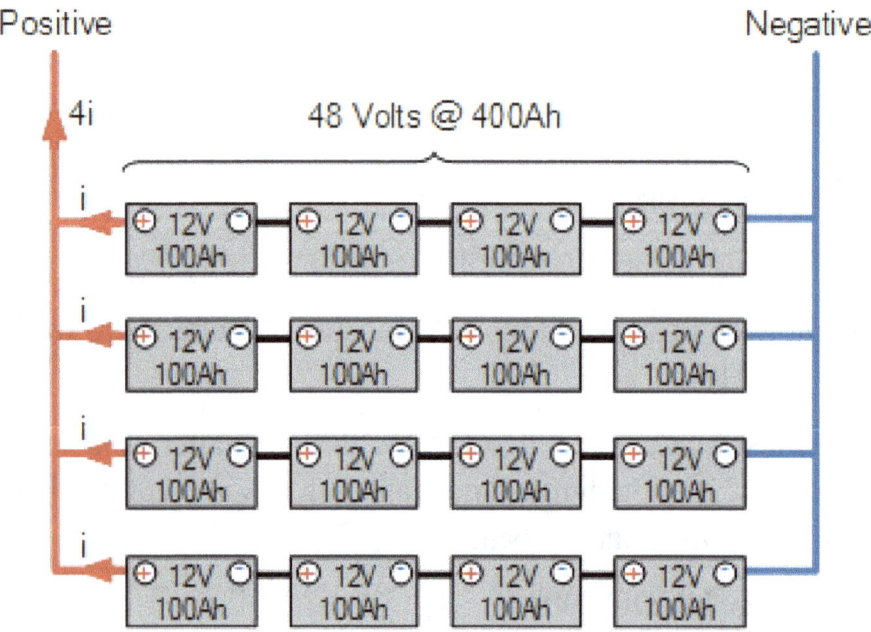

Figure 4.27 Series and parallel circuit connection

CHAPTER FIVE

5.1. Charge Controller

A charge controller or charge regulator is basically a voltage and current regulator to keep batteries from overcharging. It regulates the voltage and current coming from the solar panels going to the battery. 100W solar panels put out about 16 to 20 volts, so if there is no regulator the batteries will be damaged from overcharging. Most batteries need around 14 to 14.5 volts to get fully charged.

PV charge controller senses battery voltage. When the batteries are fully charged, the controller will stop or decrease the amount of current flowing from the photovoltaic array into the battery. When the batteries are being discharged to a low level, many controllers will shut off the current flowing from the battery to the DC load.

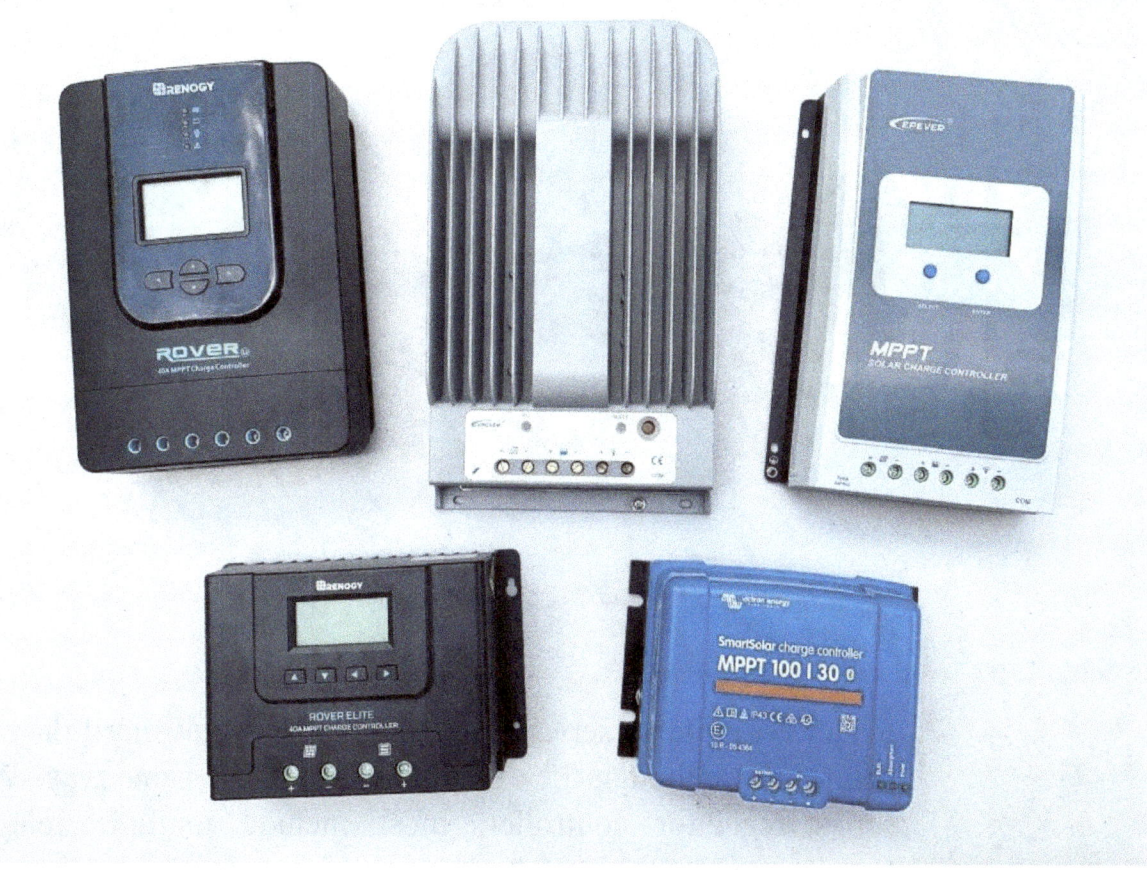

Figure 5.1 Charge controller

5.2. Types of solar charge controller

Charge controllers can be categorized into five categories.

1. **Shunt controller**: The most basic type of charge controller is a shunt controller. This operates with a basic ON or OFF switch.
When the battery voltage is low and needs to be charged, the shunt controller will go into the ON position and also when the battery is fully charged, the shunt controller will switch the OFF position and block the energy flow.

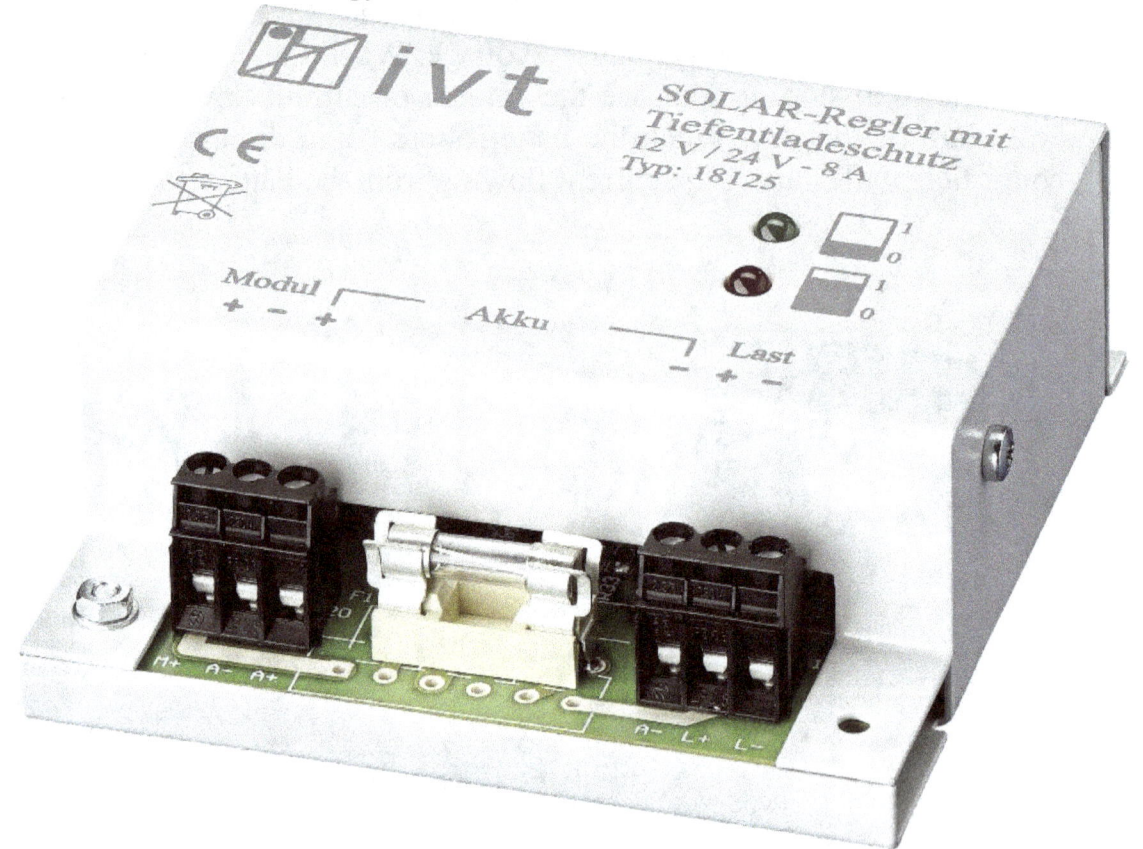

Figure 5.2 Shunt controller

2. **Series regulators:** series regulators are more advanced than shunt controllers, although they operate in much the same way. With a series regulator controller, most include multiple relays or transistors that enable multiple connection points.

3. Pulse width modulation (PWM):

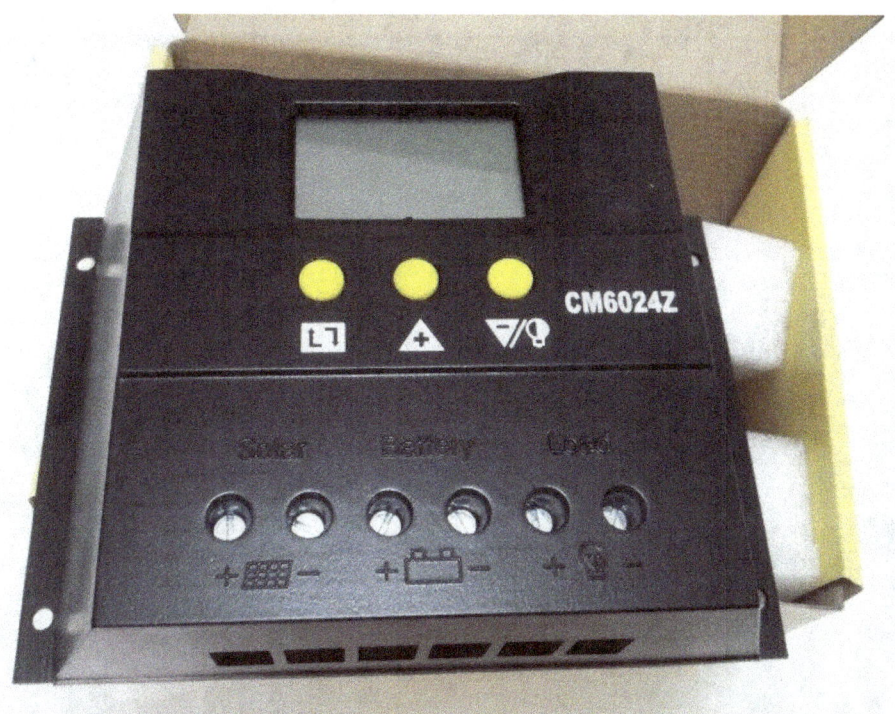

Figure 5.3 Pulse width modulation charge controller

PWM controllers enable multi-stage charging including Bulk, Absorb and Flow stages by continually regulating the energy flow. The result is a more efficient system with better charging control.

Figure 5.4 Inside a pulse width modulation charge controller (back view)

At the same time, PWM controllers are also relatively cost-effective, which is why they are recommended for occasional residential applications.

Figure 5.5 Inside a Pulse Width Modulation charge controller (front view)

4. **Maximum power point tracking (MPPT)**: One of the newest innovations for solar charge controllers, is the Maximum Power Point Tracking controllers allow solar panels to use the optimum power output voltage.

MPPT controllers are also suitable for higher voltages than the most other type of solar charge controller by allowing the best power output voltages, MPPT controllers can improve the solar panel performance by 30 percent. As the newer options on the market, MPPT controllers also have features and programming
Options.

Figure 5.6 Maximum power point tracking controller

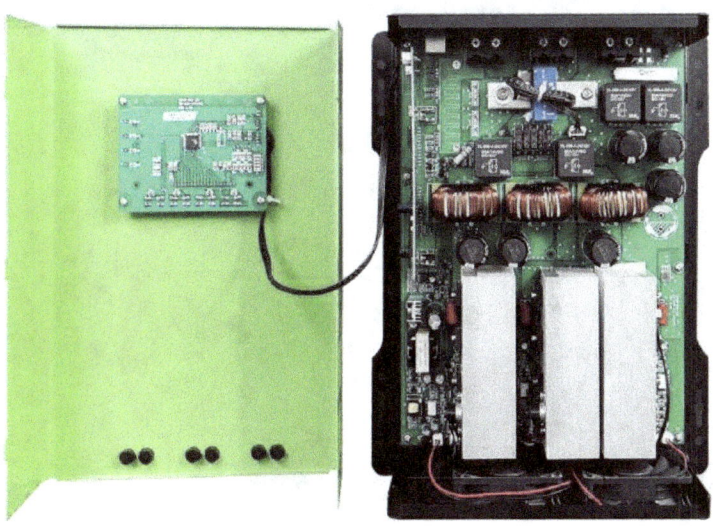

Figure 5.7 Inside maximum power point tracking controller

5. **Diversion load controllers:** Also known as a dump load controller, this works by redirecting power back to the solar array source once the battery is fully charged. Diversion load controllers are not common for residential use but are often used with wind-electric.

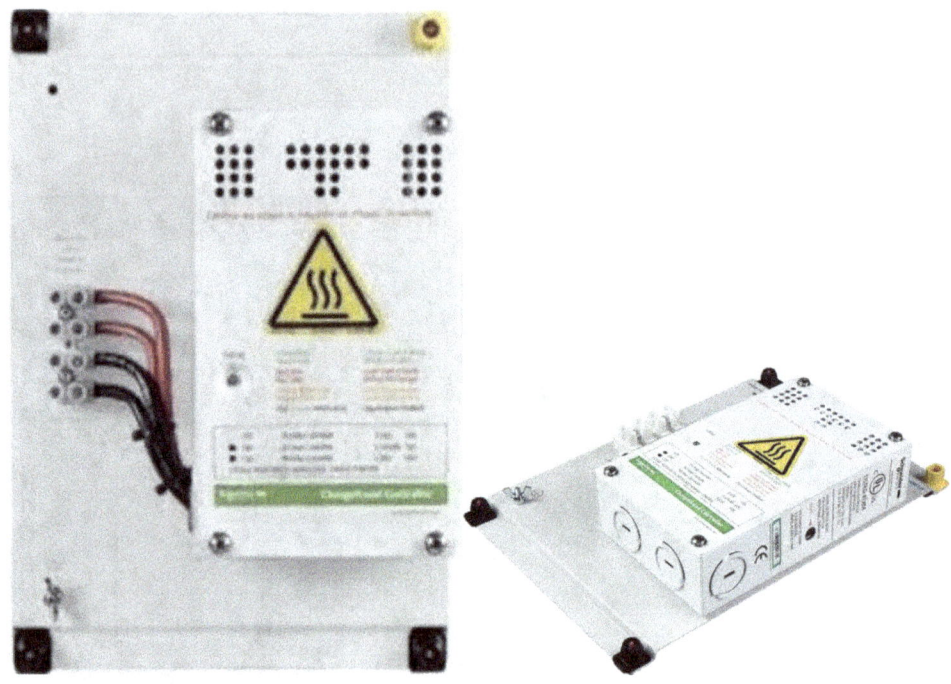

Figure 5.7 Diversion load controllers

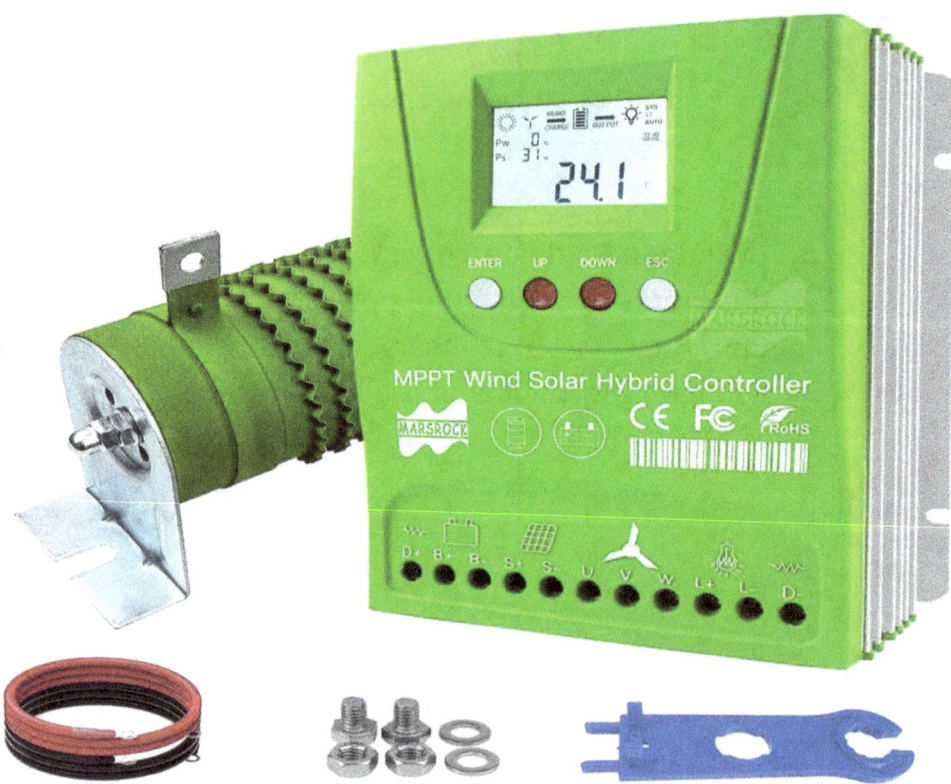

Figure 5.8 Diversion load controllers

5.3. Controller features

In addition to preventing overcharging, controllers can have many other features that protect the batteries, provide a better user interface, and increase the flexibility of a PV system.

1. **Low voltage disconnect**: Controllers can include a Low Voltage Disconnect (LVD) which disconnects the load from the battery in the event that the battery falls to a low state of charge. This feature can prevent the battery from being over-discharged, which can possibly cause permanent damage to the battery.

2. **Adjustable set point**: Controllers can include adjustable set points. This can be helpful in trying to match the ideal charge settings to the battery. Depending on the controller, adjustments can be made via jumpers, trim pots, dipswitches, display, computer, etc.

3. **Over-discharge protection:** Another main function of charge controllers is to prevent batteries from being over discharged. Like a parked car with its lights left on, photovoltaic system loads can easily over drain batteries, dramatically shortening the life of the battery. Most photovoltaic systems provide protection for the battery against unmanaged discharge.

4. **Temporarily turning off loads:** Turning off loads to prevent further battery discharge (until the photovoltaic modules or other power source recharge the battery to a minimum level) is called load management or load shedding and is accomplished using a low voltage disconnect (LVD) circuitry. If a controller performs load management, DC loads will automatically be shut down. Therefore, essential loads must be wired directly to the battery to avoid unplanned disconnection. In this case, battery over discharging can still occur since the controller has been bypassed. It is also important to remember that charge controllers only control DC loads. The inverter low voltage disconnect (LVD) needs to be programmed to disconnect the AC loads when need arises.

5. **Activating lights or buzzers:** Lights or buzzers may also be used to indicate low battery voltage and prevent cutting off critical loads. If the system is designed for critical loads, such as a vaccine refrigerator in a rural health clinic, warning lights or buzzers might be essential. However, since loads can

keep running after the user is warned, there is always the risk of over-discharging and shortening battery life.

6. **Turning on a standby power supply:** Stand-by or auxiliary power sources, such as generators, can be used to prevent over discharging. Some controllers automatically start the backup power source to recharge the battery bank when the batteries reach a low charge state. When the batteries are fully charged, the controller turns off the auxiliary power sources, and the photovoltaic system resumes it's charging operation. In "grid connected" systems with battery backup, the conventional utility power creates a backup power supply.

7. **Specifying a controller:** A photovoltaic system controller must match the system voltage. For example, a 12 volt controller is used in a 12 volt system and a 24 volt controller in a 24 volt system. Secondly, a controller must be capable of handling the maximum load current (amperage) that will pass through the controller. System designers should note that some loads might operate directly from the batteries or through an inverter and not pass through the controller. Thirdly, a controller must be able to handle the maximum PV array current. Use the maximum array amps at short circuit current (which is greater than the operating amps) plus a 25 percent safety margin to conservatively determine this figure.

   Although there are numerous optional features, system designers should consider using controllers with these features.

8. **Lights:** Indicator lights can tell users and service people how the system is operating.

   Lights can indicate when the batteries are fully charged, when the battery voltage is low, or when the LVD has shut off the loads.

9. **Temperature compensation:** when battery temperature is less than 15 degrees Celsius (59 degrees Fahrenheit) or more than 35 degrees Celsius (95 degrees Fahrenheit), the charging voltage should be adjusted. Some controllers have a temperature compensation sensor to automatically change charging voltage.

5.4. Trouble shooting a charge controller

5.4.1 Pulse width modulation (PWM):

ERROR CODE	MEANING	EXPLANATION	POSSIBLE SOLUTION
E01	Battery low or battery over discharge.	Error code (E01) will be displayed on the screen of your PWM charge controller when the battery voltage is too low and the charge controller can't charge the battery.	Boost charge the battery using an external AC battery charger.
E02	Battery over voltage.	Error code (E02) will be displayed on the screen of your PWM charge controller when the maximum battery voltage has been exceeded.	Read the manual to know the required battery voltage.
E04	DC Load short circuit.	Error code (E04) will be displayed on the screen of your PWM charge controller when there is a Short circuit in the DC load wiring connected to the charge controller.	Check the connection of the DC load connected to the charge controller.
E05	Overload of DC load.	Error code (E05) will be displayed on the screen of your PWM charge controller when the DC loads connected to the charge controller is too high.	Reduce the DC loads connected to the charge controller.
E06	Overheating.	Error code (E06) will be displayed on the screen of your PWM charge controller when the charge controller is overheating.	Check the cooling fan of the charge controller.
E08	Solar over amperage.	Error code (E08) will be displayed on the screen of your PWM charge controller when the input current from the solar array has been exceeded.	Read the manual of the PWM charge controller to know the maximum PV current and re-configure the solar array to meet the charge controller requirement.
E10	Solar over voltage	Error code (E10) will be displayed on the screen of your PWM charge controller when input voltage of the solar array has been exceeded.	Read the manual of the PWM charge controller to know the maximum PV voltage and re-configure the solar array to meet the charge controller requirement.

E13	Solar reversed polarity	Error code (E13) will be displayed on the screen of your PWM charge controller when the solar polarities has been interchanged. I.e. positive (+VE) terminal is interchanged for negative terminal (-VE) vise-vasa.	Check the solar array connections and ensure the polarities are correct and not interchanged.
E14	Battery reversed polarity	Error code (E14) will be displayed on the screen of your PWM charge controller when the battery polarities has been interchanged. I.e. positive (+VE) terminal is interchanged for negative terminal (-VE) vise-vasa.	Check the battery connections and ensure the polarities are correct and not interchanged.

Table 4.3 Pulse width modulation (PWM) error codes

5.4.2 Maximum power point tracking (MPPT)

ERROR CODE	MEANING	EXPLANATION	POSSIBLE SOLUTION
E01	Short circuit.	Error code (E01) will be displayed on the screen of your MPPT charge controller when there is a Short circuit in the DC load wiring connected to the charge controller.	Check the connection of the DC load connected to the charge controller.
E02	Over current.	Error code (E02) will be displayed on the screen of your MPPT charge controller when the input current from the solar array has been exceeded.	Read the manual of the MPPT charge controller to know the maximum PV current and re-configure the solar array to meet the charge controller requirement.
E03	Low voltage	Error code (E03) will be displayed on the screen of your MPPT charge controller when the battery voltage is too low and the charge controller can't charge the battery.	Boost charge the battery using an external AC battery charger.
E04	Battery over voltage	Error code (E04) will be displayed on the screen of your MPPT charge controller when the maximum battery voltage has been exceeded.	Read the manual to know the required battery voltage.
E05	Over temperature	Error code (E05) will be displayed on the screen of your MPPT charge controller when the charge controller is overheating.	Check the cooling fan of the charge controller.

Table 4.4 Maximum power point tracking (MPPT) error code

5.5 Tips on charge controllers

Icon	Meaning	Icon	Meaning	Icon	Meaning
 ☀	Day	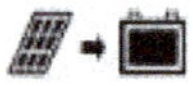 ▦→▭	Data Relates to Charging	FLT	Float Charging
 ☾	Night	 ▭→💡	Data Relates to Discharging	ABS	Absorption Charging
 ↷	Charging	 ▭ ｜	Data Relates to Temperature	EQU	Equalizing Charging
 ↱	No Charging	 ⚙	Data Adjustable	SCI	Max Charging Current
 ⤻	Load On	 ⚠⚙	Data not Adjustable	RCV	Recovery Charging Voltage
 ⤵	Load Off	 SLD	Sealed Battery	SCV	Constant Charging Voltage
 🙂	System Works Normally	 GEL	GEL Battery	LVD	Low Voltage Disconnection Voltage
 ☹	System Works Abnormally	 FLD	Flooded Battery	LVR	Low Voltage Re-connection Voltage

Figure 5.9 Tips on charge controllers

Status	Fault indicator	charging indicator	Icon	Description
Battery over discharged	Red on solid	———	⚠ 🔋	Battery level shows empty, battery frame blink, fault icon blink.
Battery over voltage	Red slowly flashing	———	 ⚠ 🔋	Battery level shows full, battery frame blink, fault icon blink.
Battery over temperature	Red slowly flashing	———	 ⚠ 🔋	Battery level shows current value, battery frame blink, fault icon blink.
Controller over temperature	Red slowly flashing	Green slowly flashing	 ⚠ 🔋	Battery level shows current value, battery frame blink, fault icon blink.
System voltage error	Red slowly flashing	Green fast flashing	 ⚠ 🔋	Battery level shows current value, battery frame blink, fault icon blink.

Figure 5.10 Tips on charge controllers

➢ Status Description

Item	Icon	Status
PV array	 ☀️▦	Day
	🌙	Night
	 ☀️▦ 🔋	No charging
	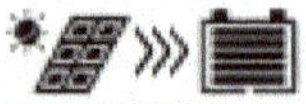 ☀️▦ »» 🔋	Charging
	PV	PV Voltage, Current, Power
Battery	 🔋	Battery capacity, In Charging
	BATT.	Battery voltage, current, temperature
	BATT. TYPE	Battery type
Load	 💡	Load ON
	 💡	Load OFF
	LOAD	Load Voltage, Current, Load mode

Figure 5.11 Tips on charge controllers

Status Icon	Indication	Status	Description
 ⬛➡	Solar Charge Indication	Flowing	Solar Power Charging Battery
		Off	Solar Power Not Charging Battery
 ➡💡	DC Load Indication	Flowing	DC Load Drawing Power
		Off	DC Load Off
MPPT	Charge Mode	Steady On	MPPT Charge Mode
BOOST			Boost Charge Mode
FLOAT			Float Charge Mode
		Off	Not Charging
CHG_V	Voltage Setting	On	Setting Charge Voltage
		Off	Charge Voltage Has Been Set
LDV_V	Over Discharge Volt Settings	On	Setting Discharge Voltage
		Off	Discharge Voltage Has Been Set
 ⬛	Solar Icon	Steady On	Daylight Detected
		Off	No Daylight Detected
		Fast Flash	Solar System Over Voltage
 🔋	Battery Icon	Steady On	Battery Connected and Functional
		Off	No Battery Connection
		Fast Flash	Battery Over-Discharged
 💡	Load Status	Flash	DC Load Short Circuit or Over-Load
		ON	Load On
		OFF	Load Off

Figure 5.12 Tips on charge controllers

Item	Icon	Status
PV array	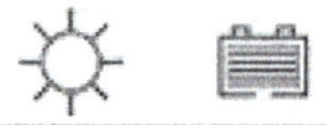 ☼ 🔋	During the day, not charging
	 ☼ ⇒ 🔋	During the day, charging
	☾	Night
	PV	PV voltage, current and ampere hours
	PV T	The total charge ampere hours of the solar panel
Battery	 🔋	Battery capacity
	⚙ BAT	Battery voltage(Programmable)
	BAT	Battery current
	BAT SOC	Battery capacity
	25 ℃	Temperature
	⚙ BAT T GEL	Battery type(Programmable)
Load	LOAD	Load voltage, current and ampere hours
	LOAD T	The total discharge ampere hours of the load
	⚙ LOAD M	Load mode(Programmable)
	 🔋 ⇒ 💡	The load is on
	🔋 💡	The load is off

Figure 5.13 Tips on charge controllers

Symbol	Meaning	Symbol	Meaning
(arrow)	load switched off	(panel→battery)	charge data
(arrow)	discharge on	(battery→bulb)	discharge data
(arrow)	charge on, flash if charge OCP	COM	communication
(arrow)	charge off	(gear)	in setup mode
(battery + thermometer)	ambient temperature	(warning + gear)	parameters not adjustable
(sun/panel)	daytime	LVD	LVD parameter
(moon/panel)	nighttime	LVR	LVR parameter
GEL	battery type: Gel	FLOAT	float charge
SLD	battery type: Sealed	ABSORB	absorption charge
FLD	battery type: Flooded	EQU	equalization charge
(battery smile)	battery is good	(bulb)	indicates load over-current
(bulb)	indicates short-circuit of load	(bulb)	load Mode 4

Figure 5.14 Tips on charge controllers

CHAPTER SIX

6.1. Inverter

An inverter converts the direct current (DC) voltage to an alternating current (AC) voltage. In most cases, the input direct current (DC) voltage is usually lower while the output alternating current (AC) is equal to the grid supply voltage of either 120VAC, or 240VAC depending on the country. The major component in the inverter is the step-up transformer which steps-up the direct current (DC) input to alternating current (AC) this is usually done by inverters with in-built step up transformer.

Figure 6.1 Inverter transformer

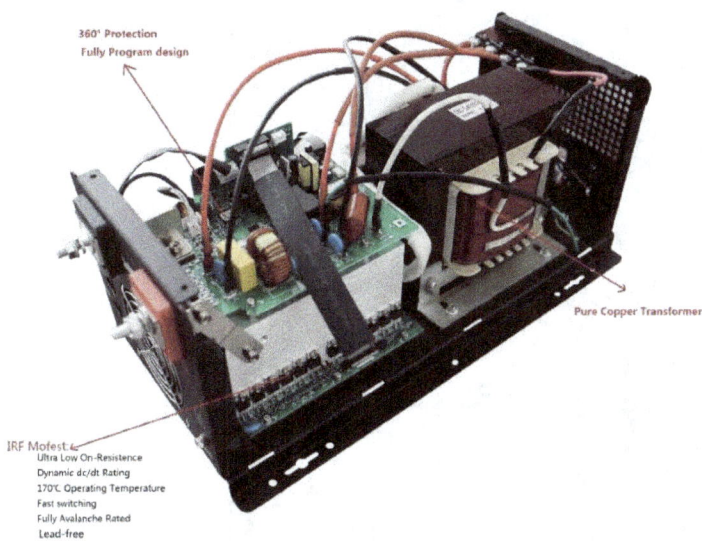

Figure 6.2 Inside transformer based inverter

But, on the other hand due to advance in technology there are inverters without transformers inside, they are called transformer-less inverters. They consist of special transformers called the **toroidal** transformer that transits the direct current (DC) voltages into alternating current (AC) inside the inverter.

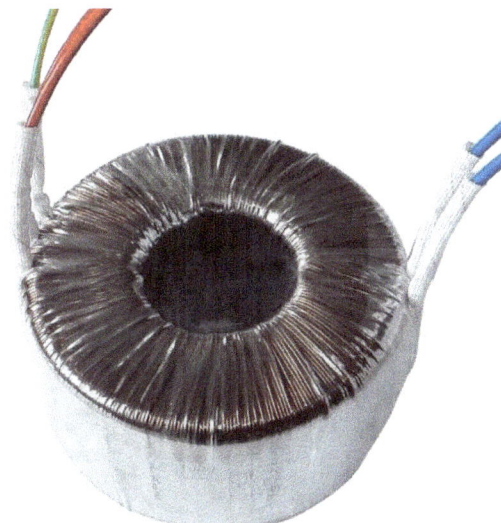

Figure 6.3 Toroidal transformer

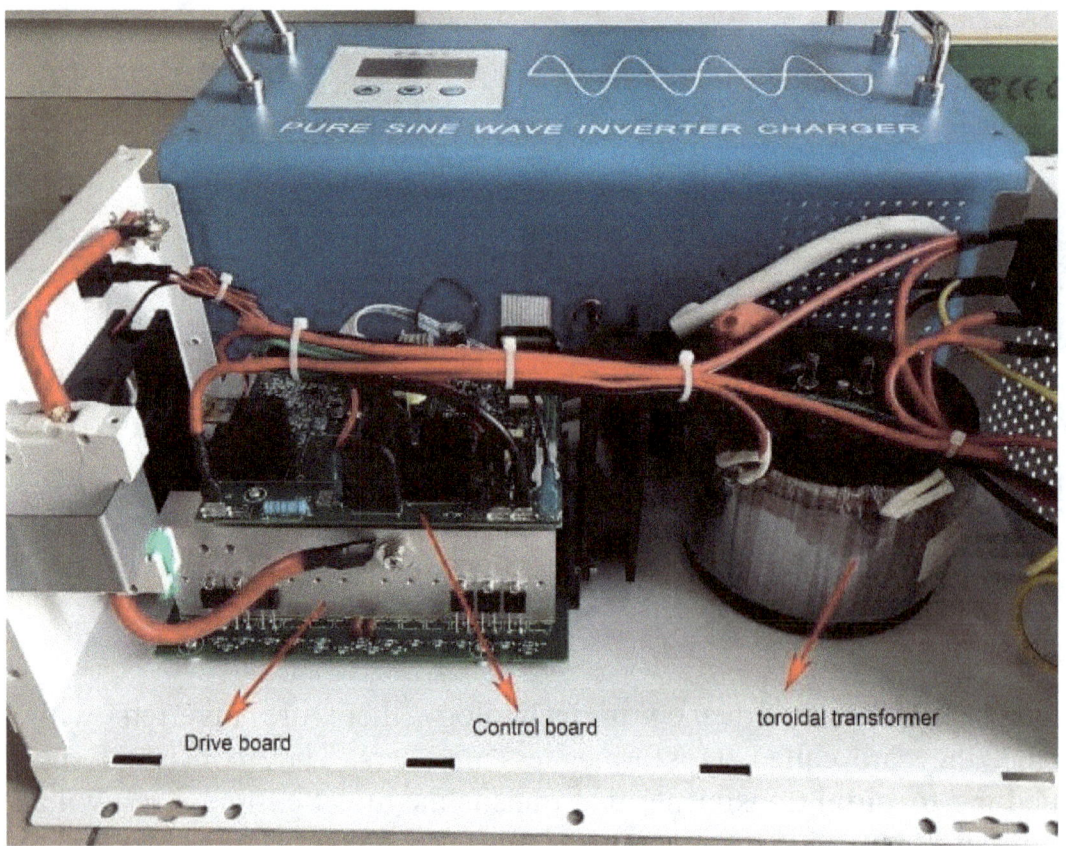

Figure 6.4 Inside a transformer less inverter

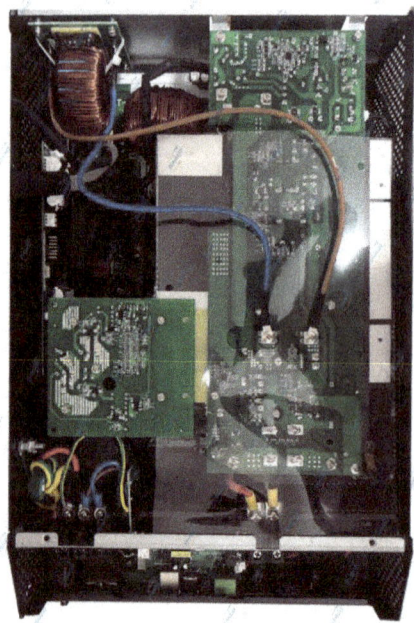

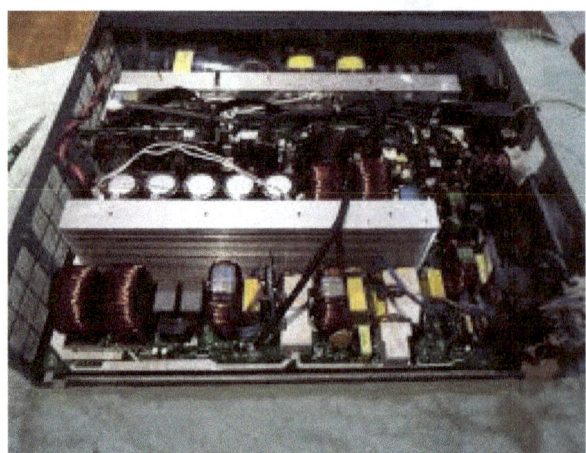

Figure 6.5 Inside a hybrid transformer-less inverter

The inverter may be built as standalone equipment for applications such as solar power, or to work as a backup power supply from batteries which are charged separately.

In a situation where there is no battery source or photovoltaic source. In this case, the inverter input DC is from the rectified mains AC in the UPS, while from either the rectified AC in the UPS when there is power, and from the batteries whenever there is a power failure.

6.2 Classification of an inverter

Inverter can be classified according to the following;

6.2.1 Classification by input source;

1. **Voltage source inverter (VSI):**
   The inverter is known as a voltage source inverter when the input is a continuous DC voltage source.
   A rigid DC voltage source with zero impedance is connected to the voltage source inverter's input. In practice, the DC voltage source's impedance is insignificant. The AC output voltage is totally dictated by the state of the switching devices in the inverter and the supplied DC power source, assuming the VSI is powered by an ideal voltage source (a very low impedance source).

2. **Current source inverter (CSI):** The current source inverter converts the input direct current into an alternating current. The current source inverter is also called current fed inverter. The output voltage of the inverter is independent of the load. The magnitude and nature of the load current depends on the nature of load impedance.

6.2.2 Classification by output phase

1. **Single phase inverter:** They are typically used in most new houses and small businesses, single phase electricity is transported via two wires, active and neutral. The electricity from the grid or your solar PV system will only flow through the one active wire, while the neutral wire is connected to the earth at the switchboard. The purpose of your home or business neutral wire is to provide a path back to your switchboard, the source of the electrical current, in the event of a fault. This will trip the breaker or blow the fuse, cutting your electricity and preventing you from getting electrocuted.

2. **Three phase inverter:** Three phase power has four wires, three of which are active, in addition to one neutral wire, which is earthed at the switchboard. Three phase electricity is common in both larger homes and businesses, and allows for smaller and less expensive wiring, and lower voltages.

6.2.3 Classification by communication technology

1. **Line communication:** The line voltage of the AC circuit is available through these inverters, and the device is turned off when the current in the SCR has a zero characteristic. Line commutation is the name given to this commutation process, and a line commutated inverter is one that works on this basis.

2. **Forced communication**: There is no zero point in the power supply with this sort of commutation. This is why the gadget must be rectified by an external source. Forced commutation is the name given to this commutation technique, and a forced commutation inverter is a device that uses it.

6.2.4 Classification by connected method

1. **Series inverter:** Inverter is an electronic circuit which converts DC power into AC power. The inverter circuit in which the commutating elements L and C are connected in series with the load to form an under damped circuit is called a series inverter.

2. **Parallel inverter**: Two thyristors, a capacitor, a center-tapped transformer, and an inductor make up a parallel inverter. An inductor is employed to maintain the current source constant while a thyristor provides a conduit for the current to flow. Commutation capacitors attached between these thyristors govern how they turn on and off.

3. **Half-bridge inverter:** An electronic circuit that enables a voltage to be applied across a load in either direction. The main idea is to use four controlled electronic switches that toggle states pair wisely.

4. **Full-bridge inverter:** The direction of current flow in the load is controlled by four controlled switches in a single-phase full- bridge inverter, The Bridge is equipped with four feedback diodes that return the energy stored in the load to the source. When all thyristors are turned off and the load isn't entirely resistive, these feedback diodes work.

5. **Three phase-bridge inverter:** Three-phase power is required for industrial and other heavy loads. A three-phase inverter is necessary to run these high loads from storage devices or other DC sources. For this, a three-phase bridge inverter can be utilized.

   Another form of bridge inverter is a three-phase bridge inverter. Which contains six controlled switches and six diodes as indicated.

6.2.5 Classification by operation mode

1. **Standalone inverter:**

   Standalone inverters are also called off grid inverters because they are not connected to the grid for charging battery. Instead the battery is charged by the photovoltaic system and the inverter draws its DC power from batteries charged by PV array and converts to AC power. Stand-alone inverter or off-grid inverter is designed for remote stand-alone application or off-grid power system with battery backup.

   Off grid or standalone inverters are mostly used in remote areas where there is no access to electricity.

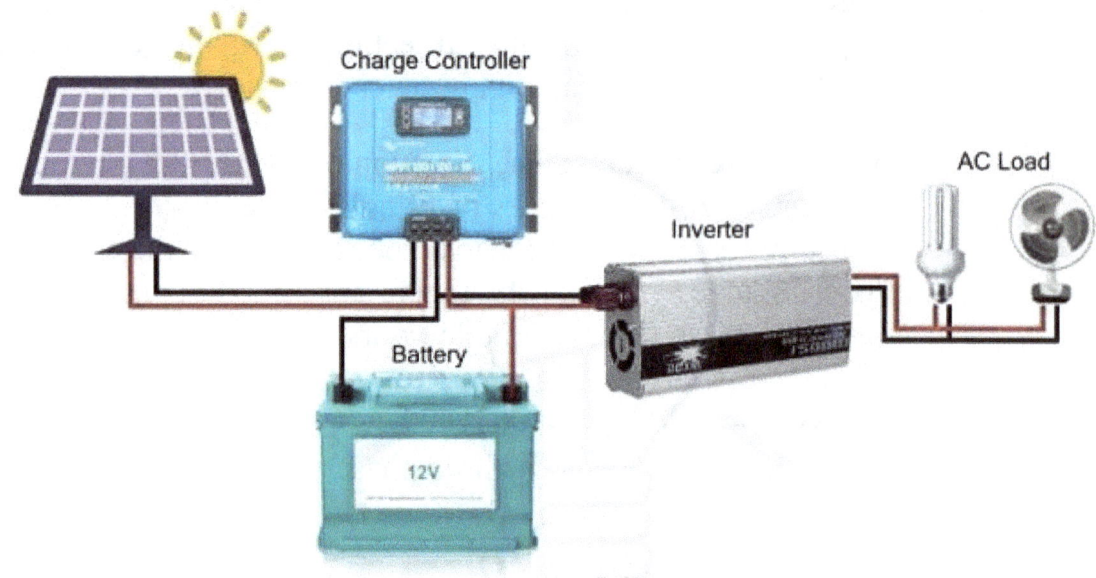

Figure 6.6 Standalone inverter system

2. **Grid inverter**

Grid inverters are kinds of inverters that gets its sources either from the grid or from the photovoltaic system. When there is no power from the grid the inverter converts the DC produced by the photovoltaic system in to AC and also when the grid power is back it stop the photovoltaic DC and by-pass the grid.

The only disadvantage of this is that it doesn't have a battery bank and it is mostly affected by weather condition.

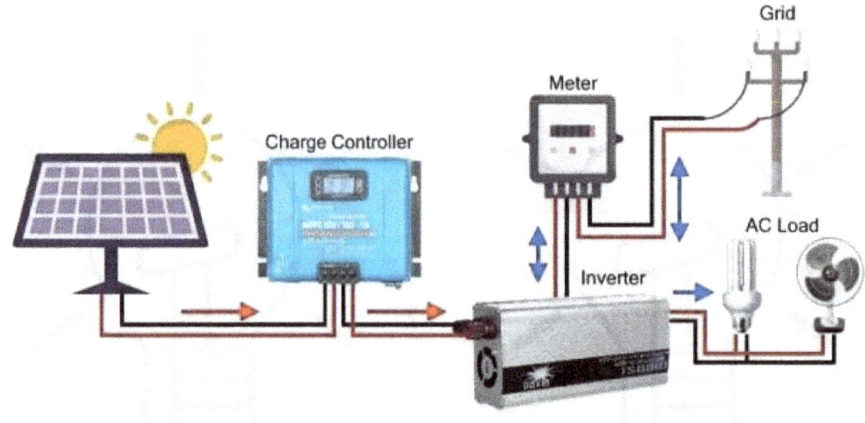

Figure 6.7 Grid inverter system

3. Bimodal inverter

The bimodal inverters is the combination of both the off grid inverter and the grid inverters, it makes use of the grid when grid power is available, the grid power can be used to charge the battery. When the grid power is not available during the day, the photovoltaic system charges the battery. And at night the battery backup can be used. The inverter converts the battery DC in to AC to power the loads at night.

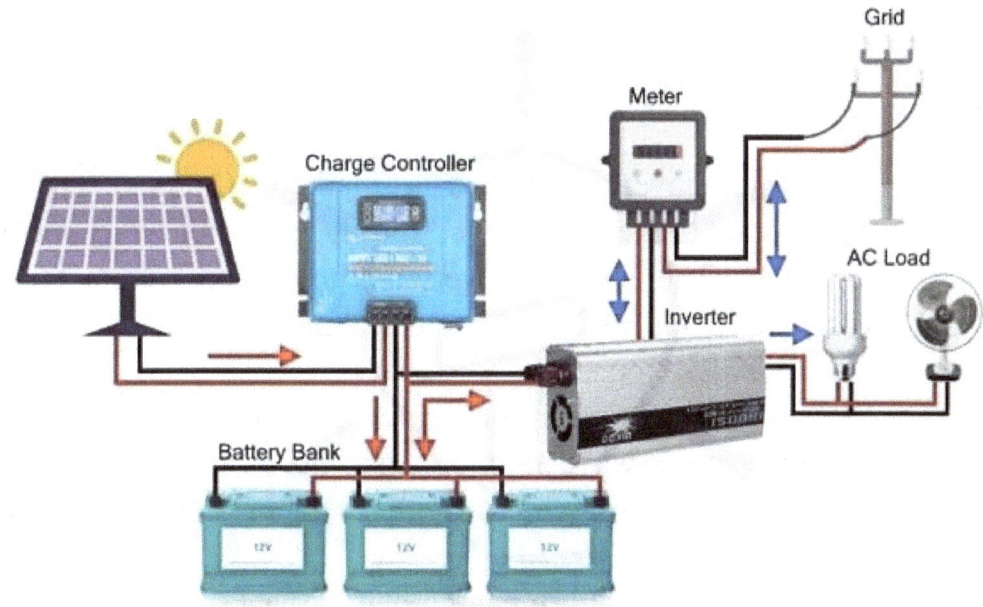

Figure 6.8 Bimodal inverter system

6.2.6 Classification by output waveform

1. Square wave inverter

These units switch the direct current input into a step-function or "square" AC output. They provide little output voltage control, limited surge capability, and considerable harmonic distortion. Consequently, square wave inverters are only appropriate for small resistive heating loads, some small appliances, and incandescent lights. These inexpensive inverters can actually burn up motors in certain equipment.

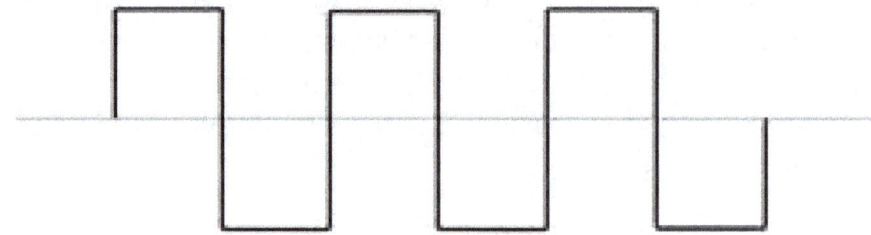

Figure 6.9 Square wave form

2.   Quasi-sine wave inverter:

This type of inverter switches DC input to AC output. These complex circuits can handle large surges and produce output with much less harmonic distortion. This style of inverter is more appropriate for operating a wide variety of loads, including motors, lights, and standard electronic equipment like TV and stereo.

However, certain electronic devices may pick up inverter noise running on a modified square wave inverter. Also, clocks and microwave ovens that run on digital timekeepers will run either fast or slow on modified square wave inverters.

It is also not advised to charge battery packs for cordless tools on modified square wave inverters.

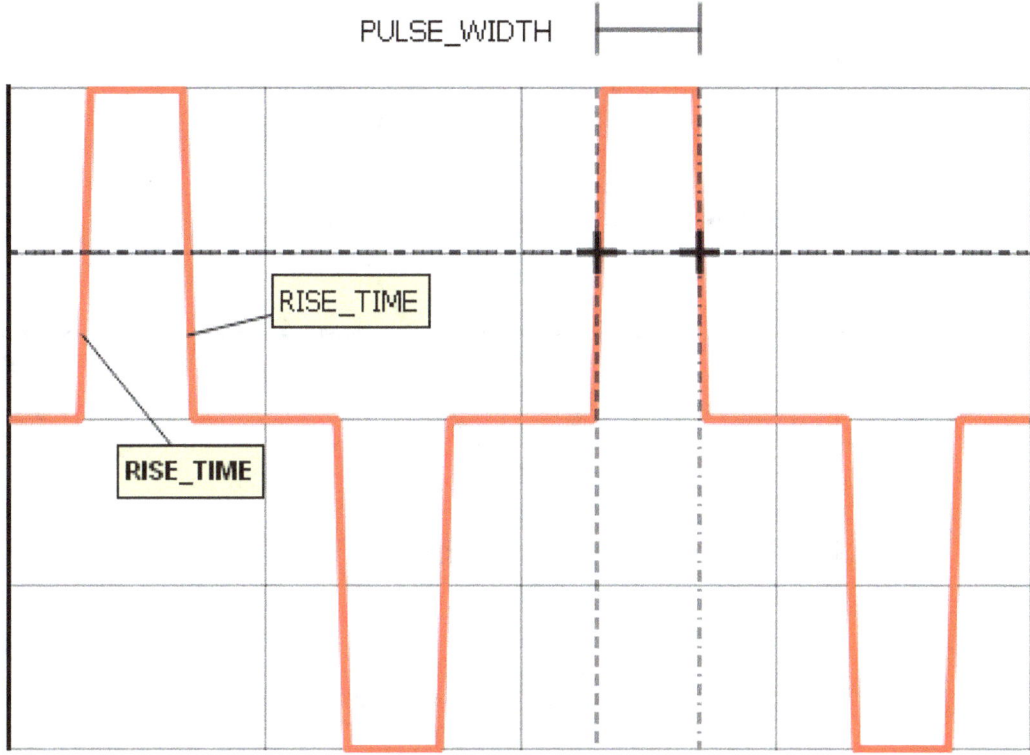

Figure 6.10 Quasi-sine wave form

3.   Pure sine wave inverter

A pure sinusoidal inverter transforms DC to AC that is almost completely sinusoidal. A pure sine wave inverter's output waveform isn't perfect, but it's far smoother than square wave and quasi-sine wave inverters.

A pure sine wave inverter's output waveform has very few harmonics. Harmonics are sine waves with variable amplitudes that are odd multiples of the fundamental frequency. Harmonics are extremely unwelcome because they can cause major issues with to appliances.

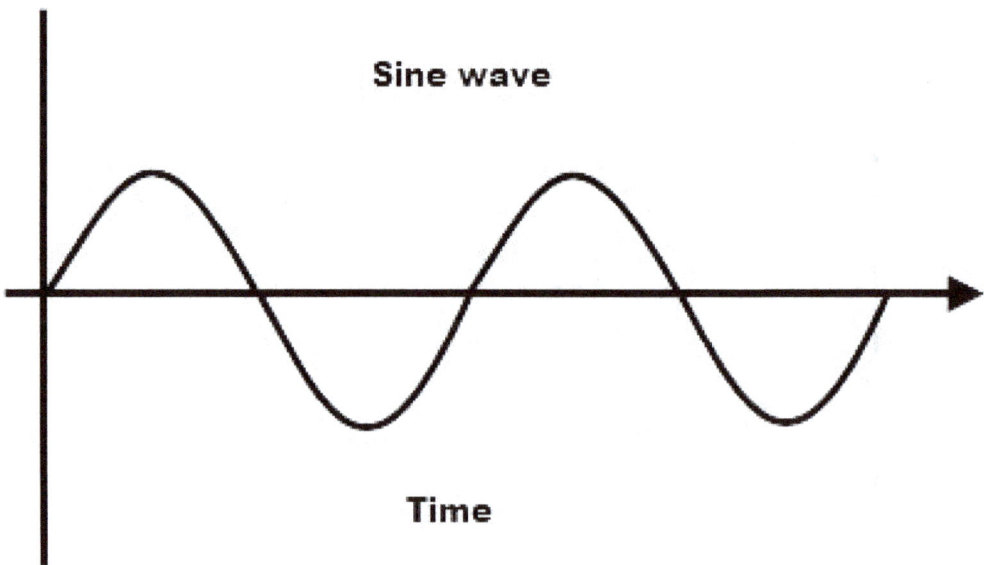

Figure 6.11 Pure sine wave form

6.2.7 Classification by number of output levels

1. **Two level inverter (TLI):** There are two output levels on a two-level inverter. At the base frequency, the output voltage cycles between positive and negative (50Hz or 60Hz).

   The output waveform of some so called "two-level inverters" has three levels. Because one of the levels in a three level inverter is zero voltage, it falls into this category. Even though zero is the third level, it is still referred to as a two-stage inverter.

2. **Multi-level inverter (MLI):** The DC signal is converted into a multilayer staircase waveform by the multilevel inverter. The multi-level inverter's output waveform is multi-level alternating rather than directly alternating positive and negative. Because the number of voltage levels is related to the smoothness of the waveform. A multilayer inverter provides a smoother waveform as a result. This characteristic, as previously stated, makes it valuable for practical applications.

6.3 Inverter Features

When buying or selecting the kind of inverter to be used, the following features must be put into consideration.

1. The inverter should convert 80 percent or more of the incoming DC input into AC output.
2. Solar inverters should be highly reliable. Currently, the solar PV power generation system is mainly applied to remote areas and remains unguarded or maintained in many cases. It requires solar inverters to be highly reliable.
3. The inverter should be highly efficient when no loads are operating.
4. Solar inverters should have a high overload capacity ranging from 125% to 150%. When the overload reaches 150%, the inverter should operate for about 30s continually. And when the overload reaches 125%, the inverter should operate for more than about 60s.
5. Solar inverters should maintain a rated sine output in any load condition except when it is overload and the transient state.
6. The inverter should provide high current required to start motors or run simultaneous loads.
7. The inverter should maintain optimum balance between the power source and load requirements.
8. The inverter should maintain 50 Hz output over a variety of input conditions.
9. Solar inverters should apply to a wide range of DC input voltage because the solar PV cell array's terminal voltage changes with the load and sunlight intensity. Although the storage battery has a clamping action on solar cells, the battery's voltage fluctuates with the remaining electricity in storage cells and changes in internal assistance. In particular, the terminal voltage changes considerably when the storage cell ages. For instance, a 12VDC storage cell has a terminal voltage ranging from 10VDC to 16VDC. Thus the solar inverter should work normally within a wide voltage range for DC input voltages. Besides, the AC input voltage should be stable and stay within the load requirements.
10. The inverter should "smooth out" unwanted output peaks to minimize harmful effects on appliances.

11. Solar inverters should have a reasonable circuit structure and consist of carefully selected components. Besides, inverters should be equipped with diverse protection functions, such as reverse polarity protection for DC input, short-circuit protection for AC output, protection against overheat and overload, etc.
12. Many PV systems have a backup AC power source, such as a generator, to charge the batteries. Battery charging capability on an inverter allows the generator to charge the batteries through the inverter (by converting the AC to DC at appropriate voltage) instead of through a separate battery-charging component.

6.4. Specifying an Inverter

It is important to choose an inverter that will satisfy a system's peak load requirements. The inverter must have the capacity to handle all the AC loads that could be on at one time. For example, a system user may wish to power a 1000 watt air condition and a 500 watt microwave oven at the same time, a minimum of 1500 watts output would be required. The kind of inverter to be used is 3500VA or 3.5KVA inverter, the 1500 watts load will consume 43% of the inverter capacity. The total load in watts must not be up to 50% the inverter capacity.

1. **Input Voltage**

    A. **Battery Voltage input:** As an inverter's maximum rated AC wattage (VA) increases, its DC input voltage also increases. It is common to find 1,200VA inverters with a 12VDC battery input, whereas 2,500VA 12VDC battery input inverters are rare.
    For example, a 1,500VA inverter works with 12VDC battery input, 2,500VA inverter works with 24VDC battery input, also a 5,000VA or 10KVA inverter works with a 48VDC battery input. But in some cases we have 4,000VA or 5,000VA inverter having 24VDC battery input.

    Note: do not exceed the rated battery voltage limit, always read your manual.

    B. **Photovoltaic voltage input:** The photovoltaic voltage input depends on the design of the inverter. Either hybrid inverter with an inbuilt solar charge controller or a non-hybrid inverter with a separate solar charge controller. The charge controller of the PV system will determine the amount of PV array power, battery voltage and maximum PV voltage.

For example:
1. **For non-hybrid inverters:** this is a kind of inverter that uses an external charge controller, i.e. the charge controller is not built together with the inverter. The charge controller of the PV System will determine the amount of PV array power, battery voltage and maximum PV voltage.

MODEL	BSM-SCH-50A	BSM-SCH-60A	BSM-SCH-70A	BSM-SCH-80A	BSM-SCH-90A	BSM-SCH-100A
Solar System Voltage	12V/24V/36V/48V Auto work					
ELECTRICAL						
PV Operating Voltage	18~150Vdc@12V 35~150Vdc@24V 50~150Vdc@36V 60~150Vdc@48V					
Max. PV Input Power	12V 700W	12V 900W	12V 1000W	12V 1200W	12V 1300W	12V 1400W
	24V 1400W	24V 1800W	24V 2000W	24V 2300W	24V 2600W	24V 2800W
	36V 2200W	36V 2600W	36V 3000W	36V 3300W	36V 3800W	36V 4200W
	48V 2900W	48V 3400W	48V 4000W	48V 4600W	48V 5000W	48V 5400W
Rated Output Current	50A	60A	70A	80A	90A	100A
Rated DC Load Current	60A					
Self Consumption	2W					
Max. Conversion Efficiency	98%		98.5%			
Protection	PV array short circuit, over charging, battery reverse polarity, output short circuit					
BATTERY CHARGING						
Battery Type	Sealed Gel AGM Flooded Lithium					
Charge Algorithm	4 stage: Bulk, Absorption, Float, Equalize charging					
Bulk Charge Voltage	Sealed: 14.4V AGM 14.2V Gel: 14.2V Flooded: 14.6V User defined: 10-15V					
Float Charge Voltage	Sealed/Gel/AGM: 13.8V Flooded: 13.7V User defined: 10-15V					
Equalize Charge Coltage	Sealed 14.6V AGM: 14.8V Flooded: 14.9V					
Load Disconnect Voltage	10.5V~12V Adjustable					
Load reconnect voltage	12.5V					
Temperature Compensation	5mV/°C/2V with BTS					
COMMUNICATION						
Communication Port	RS485					
PHYSICAL						
Net Weight	6.5kg	6.5kg	7.8kg	7.8kg	9.0kg	9.0kg
Gross Weight	7.3kg	7.3kg	8.7kg	8.7kg	10kg	10kg
Dimension(mm)	420*320*215		455*320*215		485*320*215	
Cooling	Heatsink cooling					
Enclosure	IP54					
ENVIRONMENT						
Ambient operating temperature	-25°C~60°C					
Storage Temperature	-40°C~80°C					
Humidity	100% non-condensing					
Warranty	Two years					

Technical data for 12V system at 25°C, twice in 24V system rate, triple in 36V System rate and quaduple in 48V system rate

Figure 6.12 Charge controller data sheet

For example:
80A charge controller will have battery voltage of 48VDC, it can be used for 12VDC, 24VDC, 36VDC, 48VDC. But must not exceed 48VDC battery voltage.

You must take note of the following;

I. **Maximum solar input voltage:** From the controller datasheet above in figure 6.12 the maximum solar input voltage is 150VDC. This means your PV array connected in series must not exceed 150VDC.

II. **Maximum input power:** From the controller datasheet above in figure 6.12 maximum input power depends on the input battery voltage.
   - For 12VDC battery input maximum PV power 1,200W. I.e. the numbers of solar panels connected either in series or parallel must not exceed 1,200W for 12VDC battery input into the charge controller.
   - For 24VDC battery input maximum PV power 2,300W. I.e. the numbers of solar panels connected either in series or parallel must not exceed 2,300W for 24VDC battery input into the charge controller.
   - For 48VDC battery input maximum PV power 4,600W. I.e. the numbers of solar panels connected in parallel or series must not exceed 4,600W for 48VDC battery input into the charge controller.

III. **PV Array MPPT Voltage Range:** From the controller datasheet above in figure 6.12, which talks about the operating voltage i.e. maximum and minimum voltage range for a specific DC battery input.

   For example;
   - For 12VDC battery input the voltage range of the PV must be 18VDC at minimum and 150VDC at maximum. This is because 18V is higher than 12VDC for the battery to charge. If it is below this range the battery might not charge.
   - For 24VDC battery input the voltage range of the PV must be 35VDC at minimum and 150VDC at maximum. This is because 35VDC is higher than 24VDC for the battery to charge. If it is below this range the battery might not charge.
   - For 48VDC battery input the voltage range of the PV must be 60VDC at minimum and 150VDC at

maximum. This is because 60VDC is higher than 48VDC for the battery to charge. If it is below this range the battery might not charge.

Note: the voltage range can be lower that the given range and can also be higher than the given range this depends on the battery system and the type of charge controller being used, always read your manual.

2. **For hybrid inverters:** for a hybrid inverter in which the solar charge controller is inside the inverter, same procedure should be followed with more caution. Because any mistake can damage the whole system. To do this, first open the inverter manual under specifications to know the following;

MODEL	MPS-1K-24	MPS-3K-24		MPS-5K-48
Rated Power	1000VA/800W	3000VA/2400W		5000VA/4000W
INPUT				
Voltage	230 VAC			
Selectable Voltage Range	170-280 VAC (For Personal Computers) 90-280 VAC (For Home Appliances)			
Frequency Range	50 Hz/60 Hz (Auto sensing)			
OUTPUT				
AC Voltage Regulation (Batt.Mode)	230VAC ± 5%			
Surge Power	2000VA	6000VA		10000VA
Efficiency(Peak)	90%-93%	93%		
Transfer Time	10 ms (For Personal Computers) 20 ms (For Home Appliances)			
Waveform	Pure sine wave			
BATTERY & AC CHARGER				
Battery Voltage	24 VDC	24 VDC		48VDC
Floating Charge Voltage	27VDC	27 VDC		54 VDC
Overcharge Protection	31 VDC	31 VDC		60 VDC
Maximum Charge Current	10A/20A	20A or 30A		10A/20A/30A/40A/50A/60A
SOLAR CHARGER				
Maximum PV Array Power	600 W	600W	1500W	3000W
MPPT Range @ Operating Voltage	30VDC~66VDC	30VDC~66VDC	30VDC-115VDC	60VDC~115VDC
Maximum PV Array Open Circuit Voltage	75 VDC	75 VDC	145 VDC	145VDC
Maximum Charging Current	25A	25A	60A	60A/80A
Maximum Efficiency	98%			
Standby Power Consumption	2W			
PHYSICAL				
Dimension, DxWxH (mm)				
Net Weight (kgs)				
OPERATING ENVIRONMENT				
Humidity	5% to 95% Relative Humidity(Non-condensing)			
Operating Temperature	0° C-55° C			
Storage Temperature	-15° C-60° C			
Product specifications are subject to change without further notice.				
MODEL	MPS-1K-24	MPS-3K-24		MPS-5K-48
Inverter Power	800W	2400W		4000W
Pmax. generated from solar charger	25Amp 600W	25Amp 600W	60Amp 1500W	60Amp/80Amp 3000W/4000W
Best Panel Configuration	500Wp (250Wpx2pcs)	500Wp (250Wpx2pcs)	1500Wp (250Wpx6pcs)	3000Wp (250Wpx16pcs)

Figure 6.13 Hybrid inverter data sheet

I. **Battery Voltage input:** From the hybrid inverter datasheet above in figure 6.13, Read the hybrid inverter manual to know the specified battery DC voltage and work according to the guide lines given.

From the datasheet the battery voltage is 24VDC for 1KVA inverter and 48VDC for a 5KVA inverter.

Note: do not exceed the nominal battery input voltage, be very sure before connecting.

II. **Maximum solar input voltage:** read the hybrid inverter manual to know the specified maximum solar input voltage. And connect according to the guide lines given.

From the datasheet above the maximum solar input voltage is 75VDC for 1KVA inverter and 145VDC for 5KVA inverter.

Note: do not exceed the maximum solar input voltage while connecting your PV system, always read the manual.

III. **Maximum input power:** Read the manual to know the maximum input power required by the hybrid inverter.

From the datasheet maximum PV power for 1KVA inverter is 600W and 3000W for 5KVA inverter.

Note: when connecting PV systems either in series or in parallel to increase the wattage, do not exceed the maximum PV power of the hybrid inverter.

IV. **PV Array MPPT Voltage Range:** the maximum and the minimum voltage range of PV array must not be exceeded in line with the guide lines given above.

From the datasheet above the PV array MPPT voltage range which is the minimum and maximum PV array voltages. For 1KVA inverter ranges between 30VDC to 66VDC for 24VDC battery voltage and for 5KVA inverter 60VDC to 115VDC for 48VDC battery voltage.

Note: do not operate below or above the given voltage range. Operating below the voltage range will make the battery not to charge and operating above the voltage range will cause damages on the hybrid inverter.

C. Grid voltage: the grid voltage of the inverter must be between 110VAC to 240VAC this depend on the given voltage on the manual. If the input grid voltage is high the inverter will not charge the battery. And if it is too high it can damage the inverter.

3. Surge Capacity

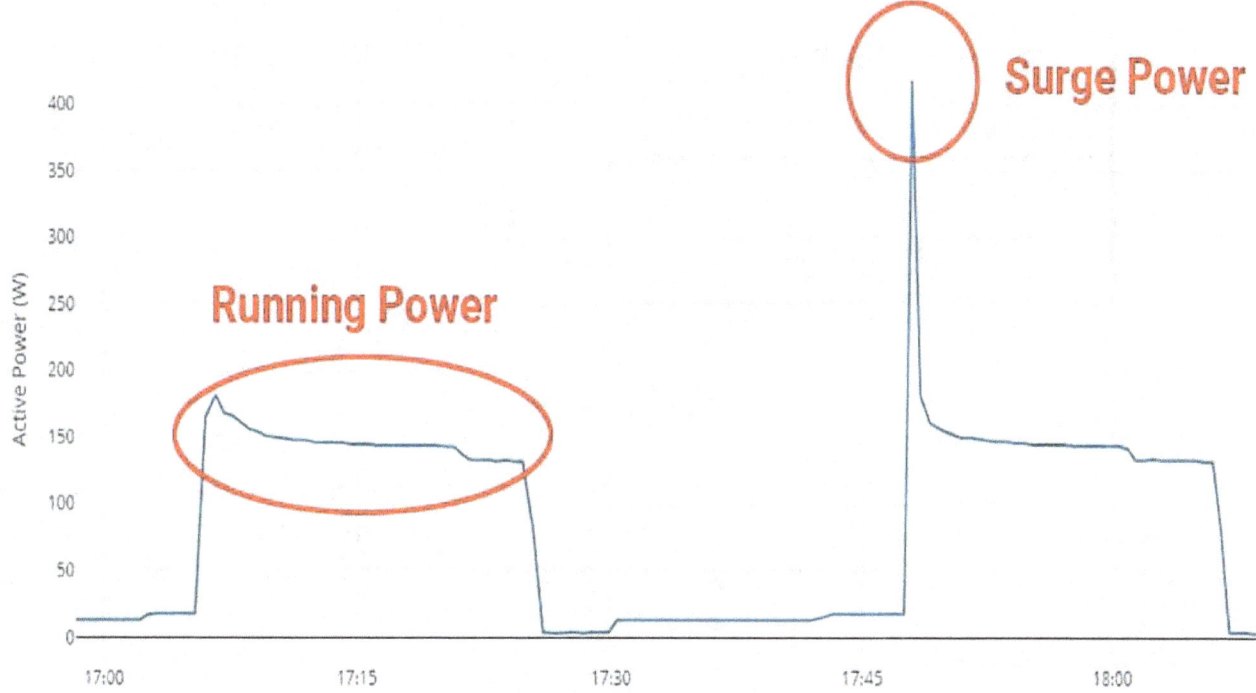

Figure 6.14 Relationship between active power and time

Most inverters are able to exceed their rated wattage for limited periods of time.

This is necessary to power motors that can draw up to seven times their rated wattage during start-up. Consult the manufacturer or measure with an ammeter to determine surge requirements of specific loads. As a rough "rule of thumb" minimum, surge requirements of a load can be calculated by multiplying the required watts by three.

For example: if a motor is rated 400W the surge requirement will be three times the rated power which is 1200W surge capacity needed to run such equipment.

4. Frequency

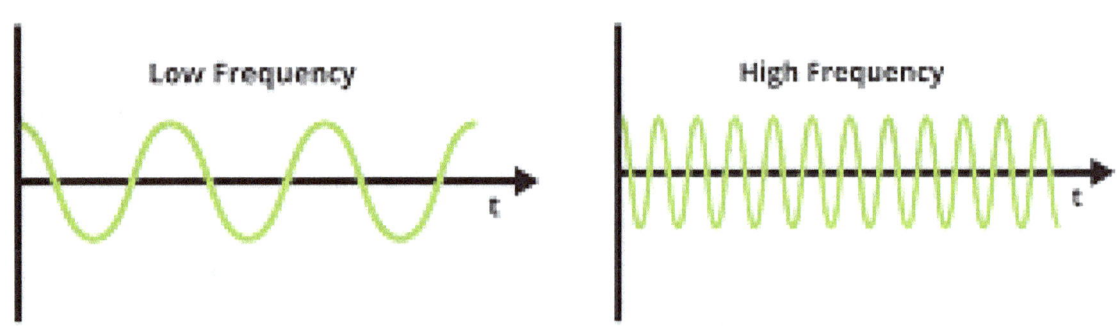

Figure 6.15 Low and high frequency wave form.

This indicates how often electricity alternates or complete a cycle. Most loads here in Nigeria uses 60 Hz. High quality equipment requires precise frequency regulation, if the input frequency of an inverter is too high, this can cause the battery not to charge.

5. Voltage Regulation
An inverter must bring out a regulated output voltage.

6. Efficiency
For an inverter to deliver high efficiency the total load must not be up to 50% the total capacity of the inverter. If it is beyond 50% going to 60% or 70% the inverter will deliver low efficiency causing the inverter to drain battery.

For example; when a 1.5KVA inverter is used to power a 1,000W load. The 1,000W load is 60.7% of the 1,500VA inverter capacity, this will reduce its efficiency by causing the inverter to drain battery.

Instead, for a 1,000W load, a 2.5KVA inverter should be used to power 1,000W load. Because the total load is 40% the inverter capacity not up to 50%. Or to archive a greater efficiency 3.5KVA inverter should be used. To have 28.5% load not up to 50% total capacity.

Note: If loaded beyond 50% or 70% of its total capacity the battery bank will drain faster. I.e. the higher the load the more battery consumption.

CHAPTER SEVEN

7.1. Sizing Photovoltaic System

Stand-alone (off grid) photovoltaic power systems are low maintenance, versatile solutions to the electric power needs of any off-grid application. They provide electric power for telecommunication stations and water pumping systems throughout the world. Twentieth-century comforts and conveniences can now be provided to remote homes via photovoltaic systems. These self-contained power stations have proven to be a reliable, cost-effective alternative to conventional power, and frequently replace the noisy, unreliable generators that most remote homes currently use.

Sizing a residential photovoltaic power system is not particularly complex. This chapter illustrates a six-step process to accurately size a system based on the user's projected needs, goals, and budget. Sizing a system includes the following steps:

- Estimating the electric load
- Sizing and specifying batteries
- Sizing and specifying an array
- Specifying a controller
- Sizing and specifying an inverter
- Sizing system wiring

This method is not fixed for a product, but rather will result in generic product specifications for the system. The method uses climatic data specific to a location and energy data specific to the user's needs.

7.2. Factor of Design

More people would have been using the solar and inverter system if not for the initial cost of building the system. But if you are looking for a better way to cut down your electricity bills the solar and inverter system is best compare to the cost of running a diesel generator. Also to reduce the cost of building the solar and inverter system one must first reduce the total load. I.e. reducing lighting load by 80% by shifting from incandescent lights to fluorescent lights (energy saving bulbs),

Figure 7.1 Fluorescent and incandescent bulb.

Changing the air conditioner to inverter compactible air conditioner, changing the fridge to inverter compactible fridge and also the television to a LCD television, this will reduce the total load of the house use and also the total cost of building the system. Because the higher the load the higher the capacity of the system needed.

In general, designers should consider the following factors when trying to optimize a system:

7.2.1 Location of installation
The site should be clear of shade to avoid loss in efficiency.

Figure 7.2 PV Site without Shade

7.2.2 Mounting Options
The optimal mounting system can maximize insolation gain.

Figure 7.3 Mounting PV system

7.2.3 Type of solar pane

The type of solar panel should be selected according to the system's parameters.

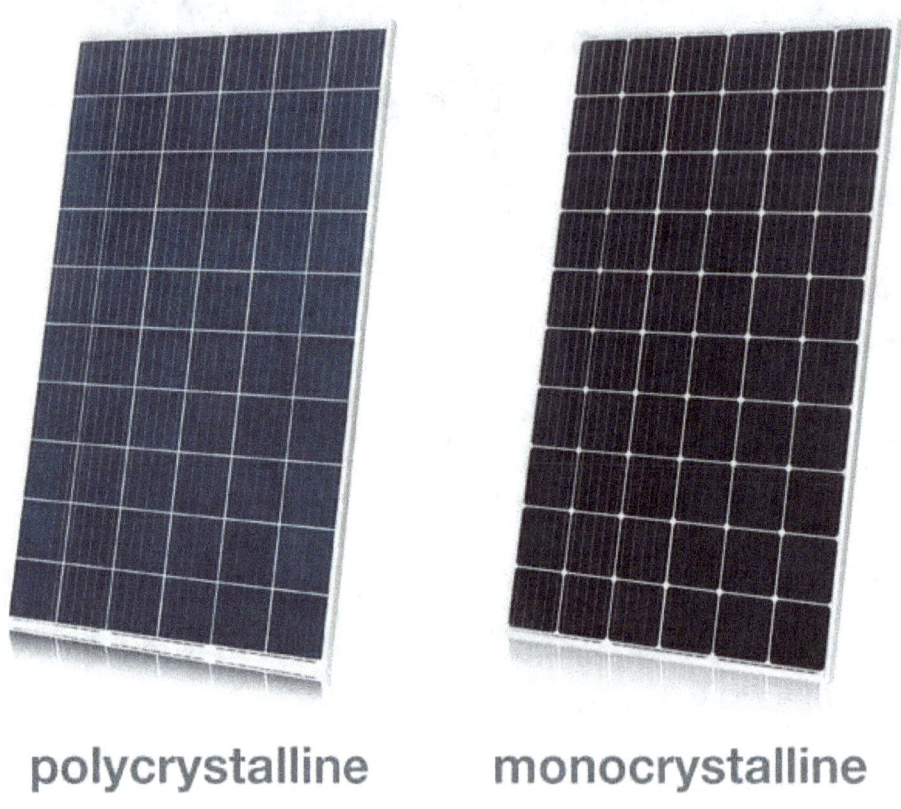

Figure 7.4 Polycrystalline and Monocrystalline solar panel

7.2.4 Wire sizing

System wiring should be designed to minimize voltage drop and provide protection from the environment. I.e. proper wire gauge should be used to prevent voltage or current loss during transmission.

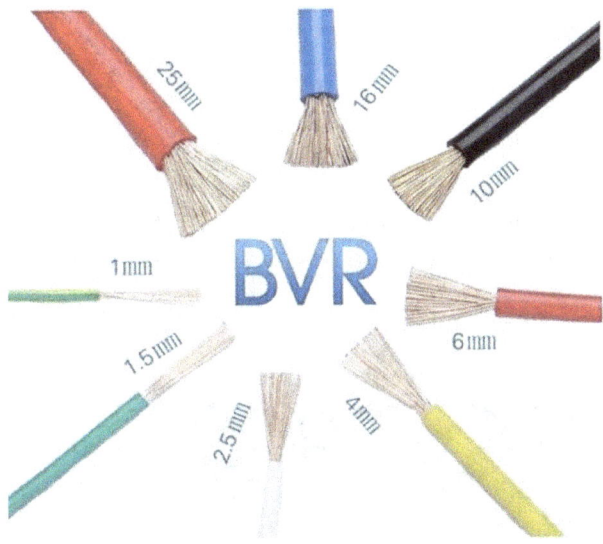

Figure 5.5 Wire sizing

7.2.5 Controllers

The charge controller must be able to carry the total power, voltage and current produced by the PV system.

MPPT

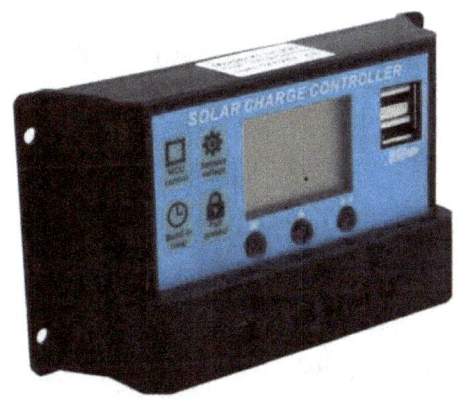

PWM

Figure 7.6 MPPT and PWM charge controller

7.2.6 Battery Storage

The battery bank must suite the PV system, charge controller, inverter and must be able to power all appliances in the house for minimum of 6 hours and 12hours depend on load application and 24hours at maximum. The type of battery to be used also must be known i.e. dry cell, tubular battery, liteum iron battery.

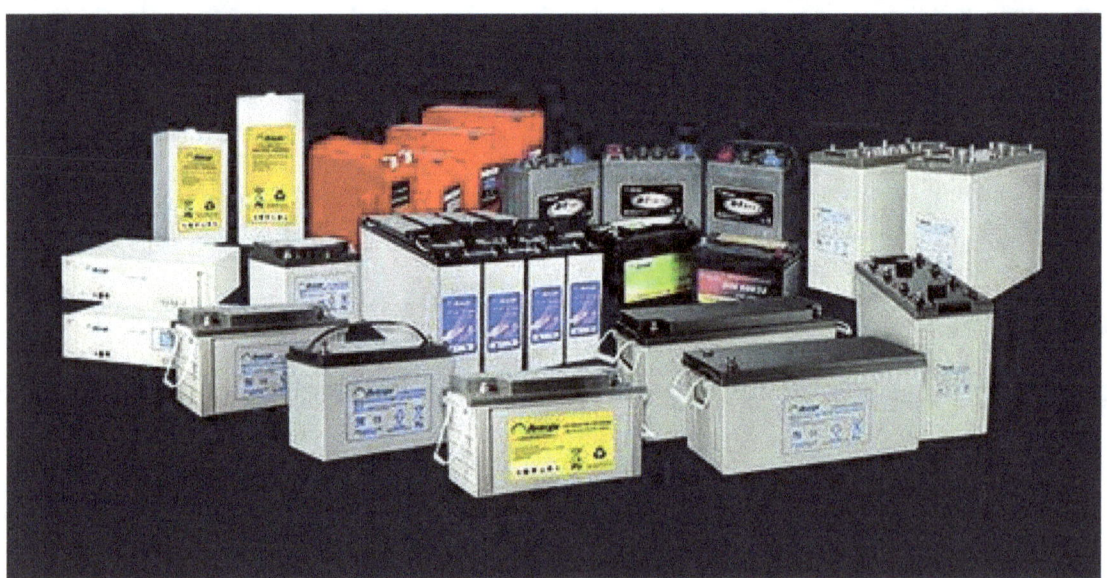

Figure 7.7 Different kinds of batteries

Also the days of autonomy must be put into consideration.

7.2.7 Loads Audit

The system loads determine the size of the system and should be minimized by intelligent planning.

7.2.8 Sizing Methodology

The photovoltaic system sizing methodology is divided into six steps that should be completed sequentially.

7.3. Electric Load Estimation

To determine the electric load estimation of a given house, you must first calculate total power consumption, this can be done by knowing the watt ratings of each equipment's to be put on the inverter system i.e. the watt rating of the laptops, television fridge, bulbs etc. all this put together will determine the size of the inverter to be used.

LOAD	POWER (W)	QUANTITY	TOTAL POWER (W)
TELEVISION	300	2	600
ENERGY SAVING BULB	6	30	180
FRIDGE	400	2	800
REFRIGERATOR	300	1	
CCTV CAMERA POWER BOX 9 CHANNELS	360	1	360
LAPTOP	200	3	600
MOBILE PHONE CHARGER	15	8	120
AIR CONDITION	1,000	1	1,000
TOTAL LOAD(W)			3,260W

Table 7.1 Load audit

From the table above, the total power in watt of all the electrical appliances to be used on inverter is 3,260W, this will determine the size of inverter to be used. We can't use 2.5 KVA, 3.5KVA and 4.5KVA inverter for this kind of system, because the total power (load) will be more than 50% of the above listed inverter capacity. Remember the higher the load, the higher the battery consumption. I.e. if we use;

- 2.5KVA (3,260W of 2,500VA) = 130.4% OVERLOAD.
- 3.5KVA (3,260W of 3,500VA) = 93.14% OVERLOAD.
- 4KVA (3,260W of 4,000VA) = 81% OVERLOAD.
- 5KVA (3,260W of 5,000VA) = 65% FAIR.
  "But, poor inverter efficiency and battery will drain faster"
- 7.5KVA (3,260W of 7,500VA) = 43% GOOD.
- 10KVA (3,260W of 10,000VA) = 32% VERY GOOD.
- 15KVA (3,260W of 15000VA) = 23.3% EXCELLENT

"More appliances can also be added for 15KVA"

Now, the inverter to be used at minimum is 7.5KVA and above, this also will determine the battery voltage to be 48VDC.

- **1.5KVA battery voltage = 12VDC**
- **2.5KVA battery voltage = 24VDC**
- **3.5KVA battery voltage = 24VDC**
- **4KVA battery voltage = 24VDC**
- **5KVA battery voltage = 48VDC**
- **7.5KVA battery voltage = 48VDC**
- **10KVA battery voltage = 48VDC**

Note: is some cases, some 4KVA inverters uses 48VDC battery and also some 5KVA inverters uses 24VDC battery this depends on the design of the inverter, always read your manual.

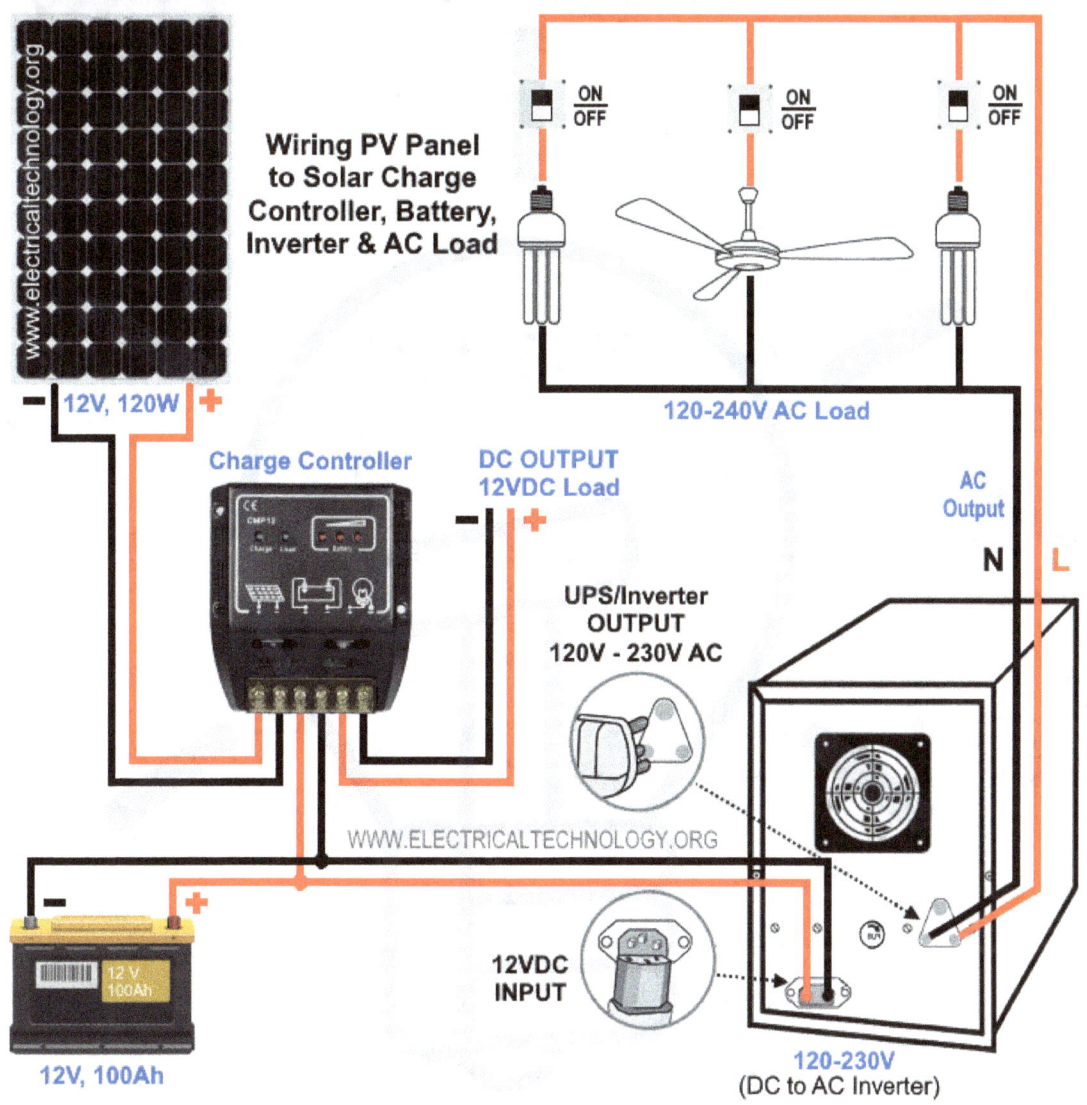

Figure 7.8 Wiring diagram of a solar and inverter system

Inverters can be connected in parallel to each other, for example, you want to build a 10KVA solar and inverter system but couldn't find the 10KVA inverter in your location, you can decide to buy two 5KVA inverters and connect them in parallel.

Connecting an inverter in parallel can be done in two ways

7.3.1 Split phase inverter configuration:
the split phase inverter is connected is such a way that the inverter P1 will have its input live wire (L1) from phase one (1) connected to the grid.

Also, inverter P2 will have its input live wire (L2) from phase two (2) connected to the grid and both inverters will have a common neutral.

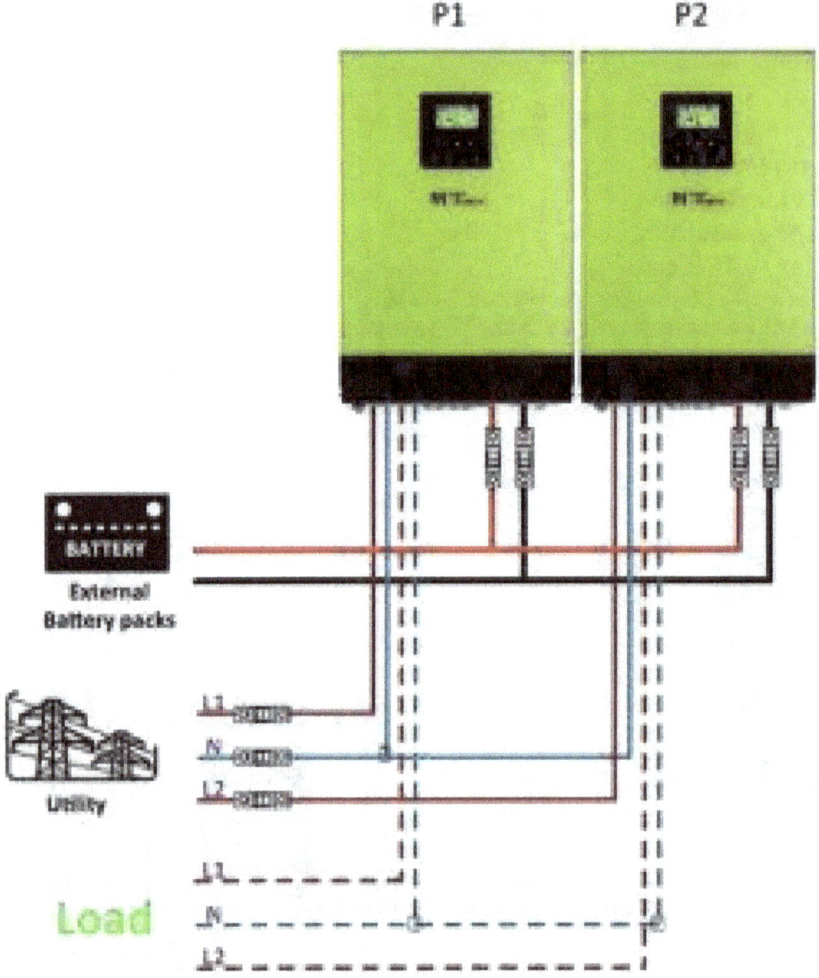

Figure 7.9 wiring diagram of a split phase inverter configuration

From the diagram above, both inverter P1 and P2 are connected to the same battery sources. The output (load) of the two inverters will also be represented as two different phases L1 output load of inverter P1 and L2 output load for inverter P2. By following the instructions in the manual set P1 as the **MASTER** and P2 as the **SLAVE**.

Note:

- *Do not connect the output load of inverter P1 and P2 together.*
- *Do not connect the current sharing cable when connecting in split phase configuration.*
- *Make sure you connect the D4 Communication cable.*
- *Both P1 and P2 inverters must be connected to the same battery bank.*

- *D4 Communication cable is for hybrid inverters only. But, for non-hybrid inverters in split phase, no need to worry about the current sharing cable and D4 communication cable. You can connect it like that.*
- *Always read the manual.*
- *All inverters must be from the same manufacturer i.e. if you buy Growatts, only Growatts must be used for all inverter in parallel, don't mix different inverters together.*

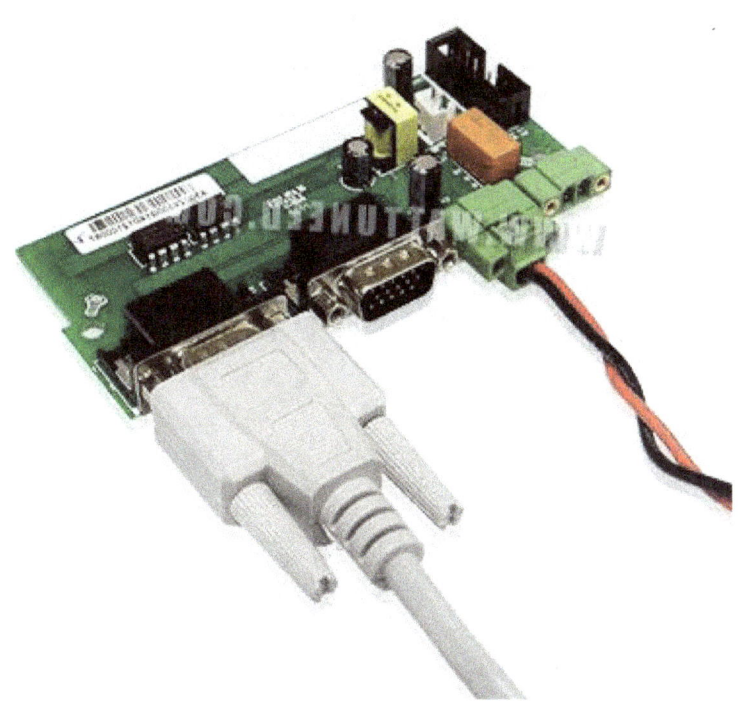

Figure 7.10 Communication cable and current caring cable in control panel

Same thing can also be done if you need 15KVA but could only get 5KVA inverter in your location. You can connect three 5KVA inverters in parallel to give you 15KVA inverter system.

For three inverters in parallel, the split phase inverter is connected is such a way that the inverter P1 will have its input live wire L1 from phase one (1) connected to the grid. Inverter P2 will have its input live wire L2 from phase two (2) connected to the grid. Also, inverter P3 will have its input live wire L3 from phase three (3) connected to the grid. All inverters will have a common neutral.

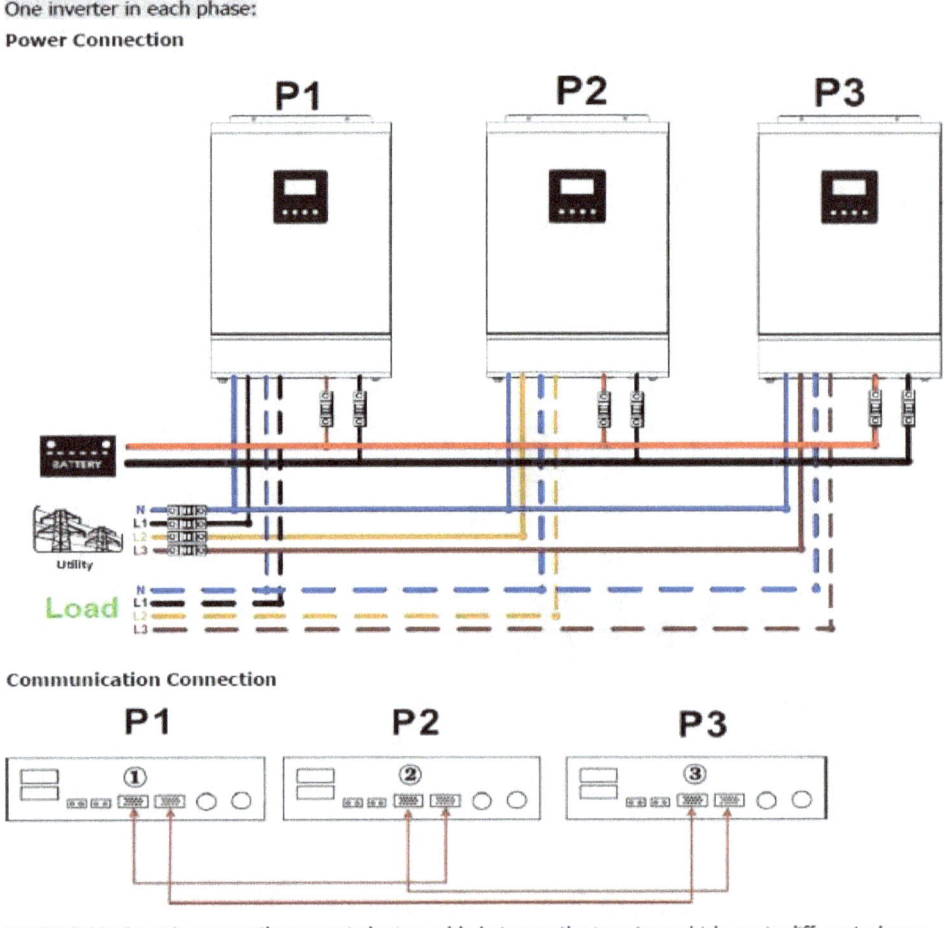

Figure 7.11 Three inverters configured in split phase

From the diagram above, both inverter P1 and P2 and P3 are connected to the same battery sources. The output (load) of the three inverters will also be represented as three different phases L1 output load of inverter P1, L2 output load for inverter P2. And L3 output load for inverter P3.

By following the instructions in the manual set P1 as the **MASTER** and P2 and P3 are the **SLAVES.**

Note:

- *Do not connect the output load of inverter P1, P2 and P3 together.*
- *Do not connect the current sharing cable when connecting in split phase configuration.*
- *Make sure you connect the D4 Communication cable.*
- *Both P1, P2 and P3 inverters must be connected to the same battery bank.*

- *D4 Communication cable is for hybrid inverters only. But, for non-hybrid inverters in split phase, no need to worry about the current sharing cable and D4 communication cable. You can connect it like that.*
- *Always read the manual.*
- *All inverters must be from the same manufacturer i.e. if you buy Growatts, only Growatts must be used for all inverters in parallel, don't mix different inverters together.*

More inverters can be connected in split phase to have a desired power ratting in KVA.

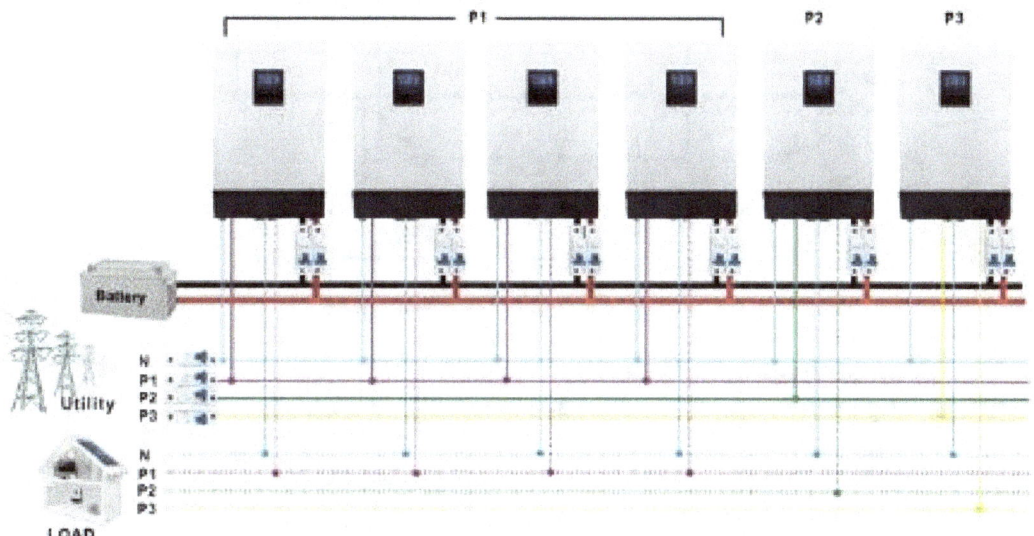

Figure 7.12 Inverter grouping configured in split phase

7.3.2 **Single phase inverter configuration:** In this kind of inverter configuration, both inverters P1 and P2 are connected to the same phase input from grid. But share the same neutral.

By following the instructions in the manual set P1 as the **MASTER** and P2 as the **SLAVE**.

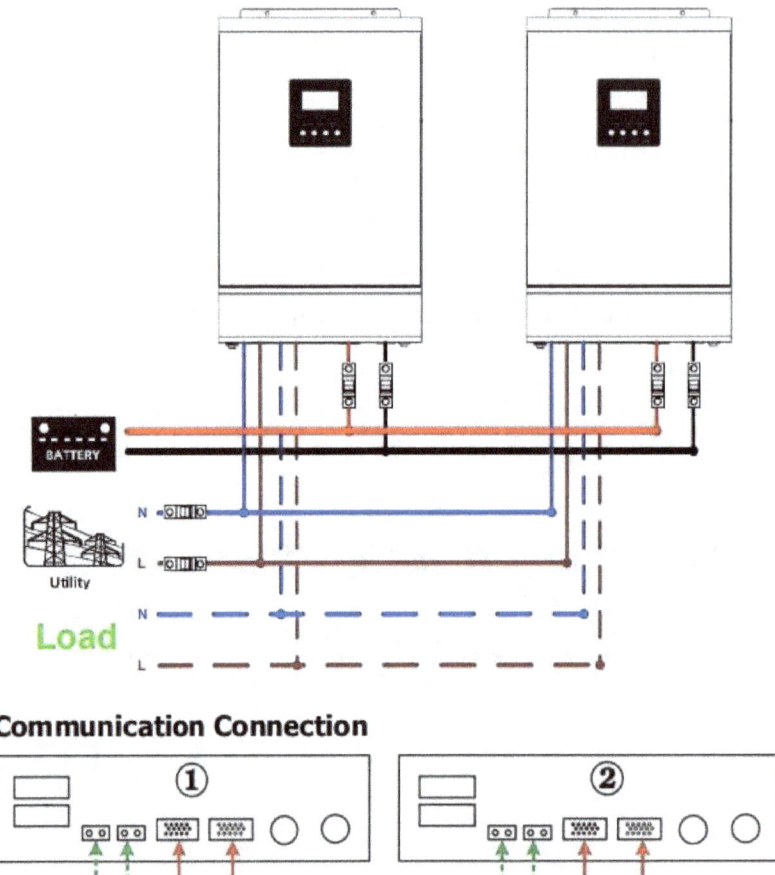

Figure 7.13 Two inverters configured in single phase

In this case the current sharing cable must be connected and also the D4 communication cable. The output of both inverters P1 and can be connected together and all inverters must be connected to the same battery.

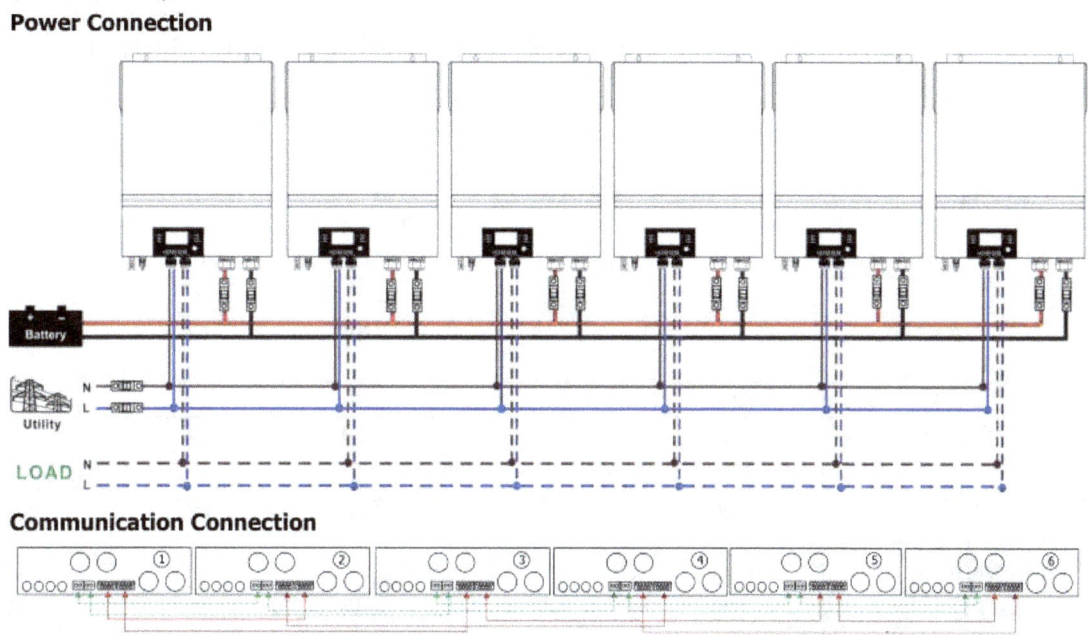

Figure 7.14 Six inverters configured in single phase

Note:
- *Connect the output load of all inverters together.*
- *Connect the current sharing cable when connecting in single phase configuration only.*
- *Make sure you connect the D4 Communication cable.*
- *All inverters must be connected to the same battery bank.*
- *This configuration is for hybrid inverters only. Do not do the same for non-hybrid inverters.*
- *Always read the manual.*
- *All inverters must be from the same manufacturer i.e. if you buy Growatts, only Growatts must be used for all inverters in parallel, don't mix different inverters together.*

7.4. Battery Sizing

To calculate the battery sizing first we must know the average daily load in watt-hour (WH)

LOAD	POWER (W)	QUANTITY	TOTAL POWER (W)	DAILY USED (HR)	AVERAGE DAILY LOAD (WH)
TELEVISION	300	2	600	8	4,800
ENERGY SAVING BULB	6	30	180	12	2,160
FREEZER	400	2	800	24	19,200
REFRIGERATOR	300	1	300	24	7,200
CCTV CAMERA POWER BOX 9 CHANNELS	360	1	360	24	
LAPTOP	200	3	600	8	8,640
MOBILE PHONE CHARGER	15	8	120	5	600
AIR CONDITION	1,000	1	1,000	8	8,000
TOTAL			3,260W		50,600WH

Table 7.2 Average daily load

Divide the watt-hours calculated in the table above by the DC system voltage to arrive at the average amp-hour per day load.

Average Amp-Hours per Day = $\frac{50,600WH}{48VDC}$ = 1,054.16 amp-hour/day

Now, to design our battery system, we must know how many batteries to put together is series and in parallel to archive 1,054.16AH battery capacity.

Let's assume to use a 12VDC, 250AH tubular battery,

No. of Batteries in Parallel = $\frac{average\ armp\ hour\ per\ day}{battery\ capacity} = \frac{1,054AH}{250AH} = 4.2$

Let's say 4 i.e. 250AH connected in parallel in 4 string.

Now to determine the numbers of batteries to be connected in series to meet up the above battery capacity.

No. of Batteries in Series = $\frac{system\ voltage}{battery\ voltage} = \frac{48VDC}{12VDC} = 4$

I.e. four 12VDC, 250AH batteries to meet up the battery capacity.

Note: this is the battery capacity needed to power the house for 1 day (24-hours), days of autonomy is not included.

To determine the numbers of batteries needed;

Multiply the numbers of battery in series and the numbers of batteries in parallel together.

Total numbers of batteries needed = 4 x 4 = 16 batteries.

See battery connection in the image below.

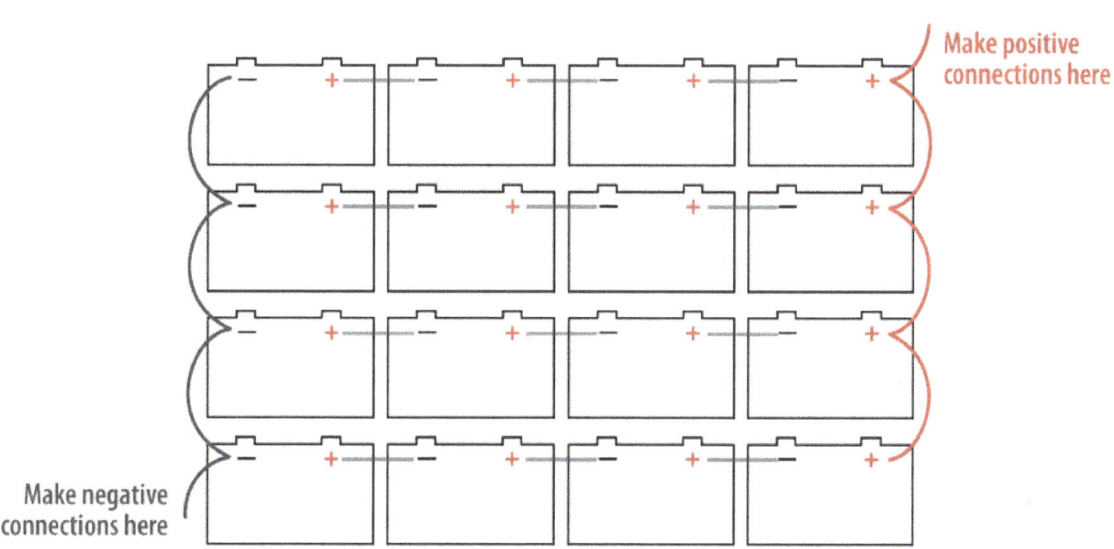

Figure 7.15 Sixteen batteries connected in series and parallel configuration to have 48VDC

Note: same must be done when calculating for the days of autonomy.

To bring in the days of autonomy which is the days of limited or no sunshine due to weather conditions such as rain. We will multiply the average amp-hour per day by 3 days depend on the design you want. You can choose 2 days if you wish but 3 days is best.

Days of autonomy = 1,054.16 amp-hour x 3days = 3,162.48AH

No. batteries in parallel = $\frac{3,162AH}{250AH} = 12$

No. batteries in series = $\frac{48VDC}{12VDC} = 4$

To determine the numbers of batteries needed;

Multiply the numbers of battery in series and the numbers of batteries in parallel together.

Total numbers of batteries needed = 12 x 4 = 48 batteries.

To determine the battery backup time:

The battery backup time is the number of hours the battery can carry the total load without the use of grid power or the photovoltaic system.

$$\text{Battery backup time} = \frac{\text{Battery capacity (AH) x Battery voltage(V)}}{\text{Total load}}$$

Four 250AH batteries are connected in parallel i.e. 250AH X 4 = 1,000AH

Battery capacity = 1,000AH.

Battery voltage = 48VDC

Total load = 3,260W

$$\text{Battery backup time} = \frac{1,000 \times 48}{3,260} = \frac{48,000}{3,260} = 14.7 = 15 \text{ hours (approximately)}$$

Battery backup time = 15 hours

This means the battery will be able to run the total load for 15 hours strait without the support of the photovoltaic system or grid power.

7.5. Sizing Your Solar Panel

Solar panels comes with different voltage and wattage sizes and are designed to supply solar power. Solar panels can be classified by their rated output power they generated, which is defined in WATTS (W).

For example, a solar panel can be ratted 200 watts peak power or 200Wp.

This wattage (W) rating is the amount of electrical power that the solar panel can produce when the sun intensity is high. And this wattage (W) can be calculated by multiplying the solar panel maximum power voltage and maximum power current.

I.e. Peak power (Pmax) = maximum power Voltage (Vmp) x maximum power current (Imp).

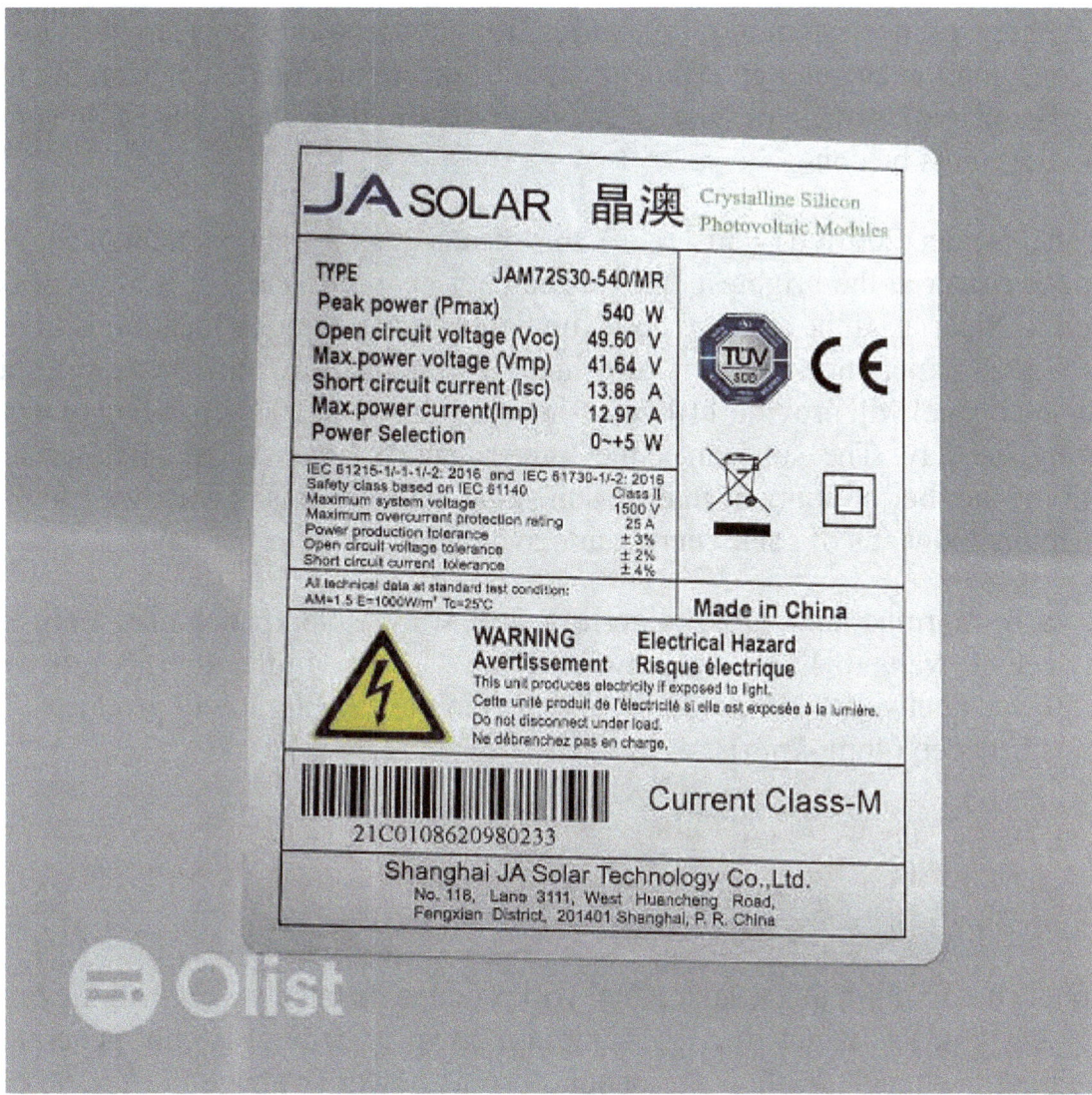

Figure 7.16 Datasheet for a 540W solar panel

From figure 7.16 by multiplying the maximum power voltage (Vmp) = 41.64VDC and maximum power current (Imp) = 12.97A, will give the peak power (Pmax) = 540W.

This can also be done manually by placing the solar panel under the sun and use a meter to measure the output voltage and current.

Multiply it together will give you the peak power of the solar panel. But accuracy will depend on the sun intensity, you can get +/-3% tolerance.

One of the problem faced by engineers is the correct sizing of the solar panel to meet the correct sizing of the system to meet the demands of the household. I.e. if few solar panels are installed it might not meet the demand of the household and also if too many solar panels are installed it might result to as waste of money and time.

The size of the photovoltaic system required varies from household to household as each house energy consumption and energy efficiency will be different. But, to determining the number of panels and total wattage of your solar system required for your household some simple mathematics must be done.

So, if a solar panel is rated 200 Wp (watts-peak), this means that the solar panel will supply 200 watts of peak power at the brightest hour of the day. I.e. when the sun is at its peak, usually 12 noon in Nigeria. If the average peak sun hours for a particular location is given as 5.5 hours (Nigeria experiences 5 hours to 6 hours of intense sun shine per day), this means that our solar panel will provide 1100 watt-hours (5.5hours x 200 Wp) a day of peak electricity during the day. The sun can shine longer than 5 hours a day, but we are concerned with the number of hours of intense sun shine, because this is what we can use to get the maximum 200 watts of solar energy, and average watts-hour per day.

For example, early morning in Lagos, Nigeria a 250 watts solar panel may only be producing between 80 watts to 120 watts between (7am - 10am), in the afternoon it will produce full 250 watts between (11am - 3pm), and also in the evening it will produce 100 watts to 60 watts between (4pm -7pm).

Note: this depends on the season and weather condition of the given location, i.e. during dry season it can be accurate but it will never be accurate during the raining season as weather may change.

Remember, people use to think that a solar panel will produce its rated wattage all the time throughout the day. But this is not true, as maximum power (wattage) is only generated when the sun intensity is high on the solar panel. Also as power (wattage) varies, so the voltage and current varies as sun intensity either increases or reduces.

there are many hundreds of different sizes of solar panels available to choose from ranging from small 100 watt panels to larger and heavier 540 watt panels at 12, 24 or 48 volts etc. and all comes with their own advantages and disadvantages. The number and type of solar panels required to capture enough solar energy to support your electrical consumption plays an important role in the design, sizing, operating voltage and cost of your solar photovoltaic system.

There are different types of solar cells to consider.

- Mono-crystalline silicon solar panels are the most efficient at converting the suns solar energy to free electricity, but they are also are the most expensive.
- Poly-crystalline silicon panels are slightly less efficient than mono-crystalline, but they tend to be cheaper since they are cheaper to produce.

- Thin film solar panels are the least efficient of them all, but they are also the cheapest and readily available

    Now, to determine the size of solar panel, we need to know the given amount of Watt-hours (or kWh) of all the electrical appliances in the house which is given as 50,600WH per day.

    By dividing this by the hour of intense sun shine or average hours of sun shine per day. i.e. $\frac{50{,}600WH}{6\ hours\ of\ sun\ shine} = 8433.33$ WH

    - To determine the numbers of solar panels needed $= \frac{8433.33WH}{540W} = 15.61 = 16$ (approximately)

    Note: you can choose any power ratting of solar panel e.g. 250W, 350W, 400W, 540W, 600W etc. but the higher the power rating of the solar panel the lesser the space occupied, a 540W solar panel will occupy less space compare to 250W solar panel.

    - To determine the numbers of solar panels to be connected in series.

Note: the numbers of solar panels to be connected in series is determined by the battery bank voltage. I.e. for 48VDC battery bank the minimum PV voltage may be 60VDC and 150VDC at maximum, this depend on the requirement of the charge controller or the PV array MPPT voltage range of the hybrid inverter, always read your manual.

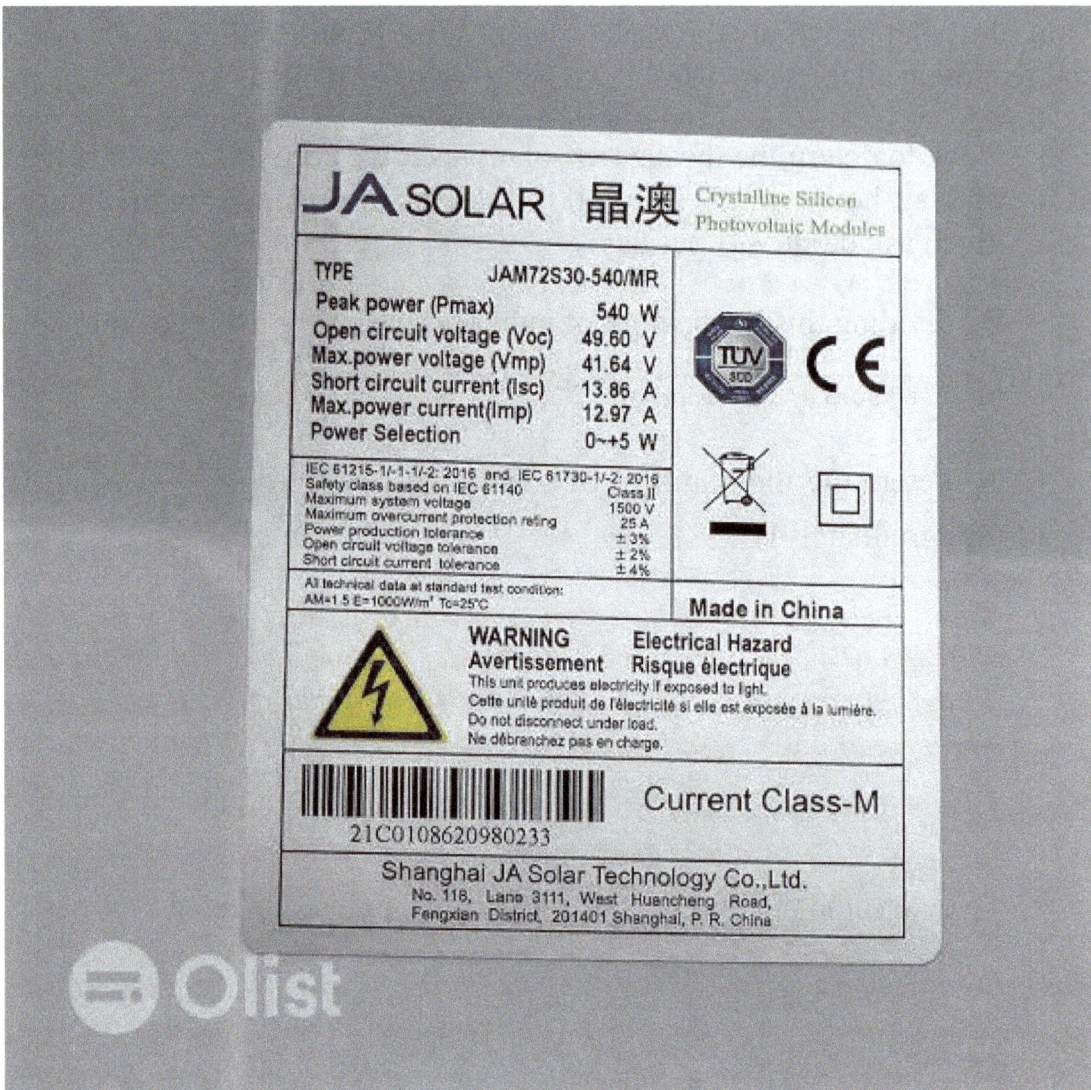

Figure 7.17 Datasheet for 540W solar panel

From figure 7.17, For a 540 watts solar panel the maximum power voltage (Vmp) is 41.64VDC, which is not up to 60VDC, the 41.64VDC cannot charge a 48VDC battery bank but by connecting two (2) in series will give 83.28VDC.

83.28VDC is greater than the minimum voltage 60VDC but, lesser than 150VDC the maximum voltage needed by the charge controller to charge the battery.

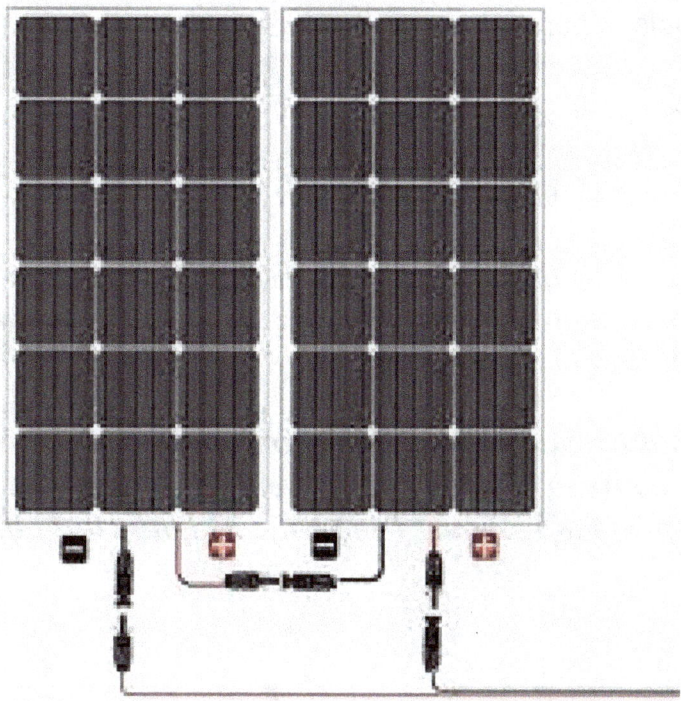

Figure 7.16 Two solar panels connected in series configuration

So, numbers of panels in series = 2

When connecting two solar panels in series, the voltage increase while the current remain the same.

I.e. (41.64 X 2) VDC = 83.28VDC

Two solar panels connected in series will produce power output of 1080.14W

I.e. power (P) = voltage (V) x current (A)

P = 83.28VDC X 12.97A

P = 1080.14W

- To determine the numbers of solar panels to be connected in series and parallel, remember that when connecting in parallel the current increase while the voltage remains the same. Also when two solar panels are connected in series the voltage increase while the current remains the same.

No. solar panels in parallel = $\frac{16}{2}$ = 8 solar panels.

Eight (8) Solar panels in parallel will produce 8,641.12W.

i.e. 1,080.14W x 8 = 8,641.12 W

"**8,641.12 W** *produced by the photovoltaic system is enough to power your load 3,260W for 6 hours to 12 hours depend on the sun intensity during the day before your battery bank continues to power at night.*

And if the battery has a backup time of 15 hours and the photovoltaic system can power the house for 6 hours to 12 hours, then we have 24 hours of electricity i.e. 15 hours of battery backup + (6 to 12)hours from PV = 21 hours to 27 hours of electricity in a day"

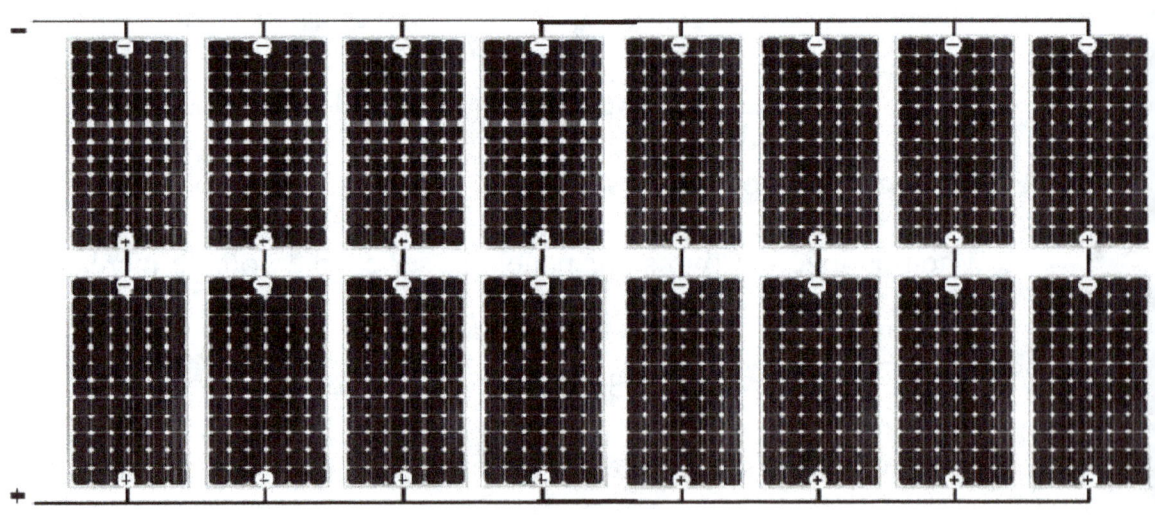

Figure 7.17 Series and parallel configuration of a solar panel

Note: the numbers of solar panels connected is parallel will determine the charge controller to be used.

So, to archive our aim we will connect the solar panel in a series and parallel configuration.

7.6. Controller Specifications

MODEL	BSM-SCH-50A	BSM-SCH-60A	BSM-SCH-70A	BSM-SCH-80A	BSM-SCH-90A	BSM-SCH-100A
Solar System Voltage	\multicolumn{6}{c}{12V/24V/36V/48V Auto work}					
ELECTRICAL						
PV Operating Voltage	18~150Vdc@12V	35~150Vdc@24V		50~150Vdc@36V	60~150Vdc@48V	
Max. PV Input Power	12V 700W	12V 900W	12V 1000W	12V 1200W	12V 1300W	12V 1400W
	24V 1400W	24V 1800W	24V 2000W	24V 2300W	24V 2600W	24V 2800W
	36V 2200W	36V 2600W	36V 3000W	36V 3300W	36V 3800W	36V 4200W
	48V 2900W	48V 3400W	48V 4000W	48V 4600W	48V 5000W	48V 5400W
Rated Output Current	50A	60A	70A	80A	90A	100A
Rated DC Load Current	60A					
Self Consumption	2W					
Max. Conversion Efficiency	98%			98.5%		
Protection	PV array short circuit, over charging, battery reverse polarity, output short circuit					
BATTERY CHARGING						
Battery Type	Sealed Gel AGM Flooded Lithium					
Charge Algorithm	4 stage: Bulk, Absorption, Float, Equalize charging					
Bulk Charge Voltage	Sealed: 14.4V AGM 14.2V Gel:14.2V Flooded:14.6V User defined:10-15V					
Float Charge Voltage	Sealed/Gel/AGM:13.8V Flooded:13.7V User defined:10-15V					
Equalize Charge Coltage	Sealed 14.6V AGM:14.8V Flooded:14.9V					
Load Disconnect Voltage	10.5V~12V Adjustable					
Load reconnect voltage	12.5V					
Temperature Compensation	5mV/°C/2V with BTS					
COMMUNICATION						
Communication Port	RS485					
PHYSICAL						
Net Weight	6.5kg	6.5kg	7.8kg	7.8kg	9.0kg	9.0kg
Gross Weight	7.3kg	7.3kg	8.7kg	8.7kg	10kg	10kg
Dimension(mm)	420*320*215		455*320*215		485*320*215	
Cooling	Heatsink cooling					
Enclosure	IP54					
ENVIRONMENT						
Ambient operating temperature	-25°C~60°C					
Storage Temperature	-40°C~80°C					
Humidity	100% non-condensing					
Warranty	Two years					

*Technical data for 12V system at 25°C, twice in 24V system rate, triple in 36V System rate and quaduple in 48V system rate

Figure 7.18 Controller specifications

Specifying the type of charge controller to be used, you can use PWM charge controller and also you can use MPPT charge controller both charge controller will charge the battery but, with different charging rate.

7.6.1. Sizing with an MPPT charge controller
Charging with an MPPT charge controller you must take note of the followings;
1. **Maximum Solar input voltages of the charge controller:**
   The maximum PV voltage of the charge controller must be well managed. Because when the input PV voltage is too high it can damage the charge controller completely and if the input PV voltage to the charge controller is too low, the battery will not charge. The maximum PV voltages determines how many solar panels to be connected in series.
   - **For 10A MPPT maximum PV voltage is 40VDC**
   - **For 20A MPPT maximum PV voltage is 100VDC**
   - **For 30A MPPT maximum PV voltage is 100VDC**
   - **For 40A MPPT maximum PV voltage is 100VDC**
   - **For 50A MPPT maximum PV voltage is 150VDC**
   - **For 60A MPPT maximum PV voltage is 150VDC**
   - **For 80A MPPT maximum PV voltage is 150VDC**
   - **For 100A MPPT maximum PV voltage is 200VDC**
   - **For 120A MPPT maximum PV voltage is 200VDC**
   - **For 130A MPPT maximum PV voltage is 200VDC**
   - **For 150A MPPT maximum PV voltage is 210VDC**
   - **For 180A MPPT maximum PV voltage is 250VDC**
   - **For 200A MPPT maximum PV voltage is 250VDC**
   - **For 220A MPPT maximum PV voltage is 260VDC**
   - **For 300A MPPT maximum PV voltage is 280VDC**

   Note: the above listed maximum PV voltage must not be exceeded or else the charge controller will get damaged.
   The charge controller maximum PV voltage might change, this depend on the design and manufacturer of the charge controller, some charge controller 80A maximum PV voltage is 115VDC, do not exceed the maximum PV voltage, always read your manual.

2. **Maximum solar input power of the charge controller.**
   The watts ratting of the charge controller must not be exceeded or else the charge controller might get damaged completely if the in rush current into the charge controller is too high.

ELECTRIFIED

The watts ratting of a charge controller depends on the amount of solar panels to be connected either in series or in parallel.

- For 10A MPPT maximum watts at 12VDC 130W/ 24VDC 150W
- For 20A MPPT maximum watts at 12VDC 260W/ 24VDC 360W
- For 30A MPPT maximum watts at 12VDC 400W/ 24VDC 800W
- For 40A MPPT maximum watts at 12VDC 480W/ 24VDC 960W
- For 50A MPPT maximum watts at 12VDC 700W/ 24VDC 1400W/ 36VDC 2200W/ 48VDC 2900W
- For 60A MPPT maximum watts at 12VDC 900W/ 24VDC 1800W/ 36VDC 2600W/ 48VDC 3400W
- For 80A MPPT maximum watts at 12VDC 1200W/ 24VDC 2300W/ 36VDC 4200W/ 48VDC 4600W
- For 100A MPPT maximum watts at 12VDC 1400W/ 24VDC 2800W/ 36VDC 3600W/ 48VDC 5400W
- For 120A MPPT maximum watts at 12VDC 1440W/ 24VDC 2880W/ 36VDC 4320W/ 48VDC 5760W
- For 130A MPPT maximum watts at 12VDC 1560W/ 24VDC 3120W/ 36VDC 4680W/ 48VDC 6240W
- For 150A MPPT maximum watts at 12VDC 1800W/ 24VDC 3600W/ 36VDC 5400W/ 48VDC 7200W
- For 160A MPPT maximum watts at 12VDC 1920W/ 24VDC 3840W/ 36VDC 5760W/ 48VDC 7680W
- For 180A MPPT maximum watts at 12VDC 2160W/ 24VDC 4320W/ 36VDC 6480W/ 48VDC 8640W
- For 200A MPPT maximum watts at 12VDC 2400W/ 24VDC 4800W/ 36VDC 7200W/ 48VDC 9600W
- For 220A MPPT maximum watts at 12VDC 2640W/ 24VDC 5,280W/ 36VDC 7,920W/ 48VDC 10,560W

Note: the above listed maximum watts must not be exceeded or else the charge controller will get damaged.
The charge controller maximum wattage might change, this depend on the design and manufacturer of the charge controller, always read your manual.

3. Maximum battery voltage

Remember, we said the battery bank voltage will determine the size of charge controller to be used.
- For 10A MPPT battery voltage is 12VDC
- For 20A MPPT battery voltage is 12VDC/24VDC
- For 30A MPPT battery voltage is 12VDC/24VDC
- For 40A MPPT battery voltage is 12VDC/24VDC/36VDC/48VDC
- For 50A MPPT battery voltage is 12VDC/24VDC/36VDC/48VDC
- For 60A MPPT battery voltage is 12VDC/24VDC/36VDC/48VDC
- For 80A MPPT battery voltage is 12VDC/24VDC/36VDC/48VDC
- For 100A MPPT battery voltage is 12VDC/24VDC/36VDC/48VDC
- For 120A MPPT battery voltage is 12VDC/24VDC/36VDC/48VDC

- For 130A MPPT battery voltage is 12VDC/24VDC/36VDC/48VDC
- For 150A MPPT battery voltage is 12VDC/24VDC/36VDC/ 48VDC / 96VDC/120VDC/192VDC
- For 180A MPPT battery voltage is 48VDC/96VDC/120VDC/192VDC/240VDC/360VDC/384VDC
- For 200A MPPT battery voltage is 96VDC/120VDC/192VDC/240VDC/360VDC/384VDC
- For 220A MPPT battery voltage is 96VDC/120VDC/192VDC/240VDC/360VDC/384VDC
- For 300A MPPT battery voltage is 96VDC/120VDC/192VDC/240VDC/360VDC/384VDC

Remember, we have high voltage and low voltage MPPT charge controllers

Note: the above listed maximum battery voltage must not be exceeded or else the charge controller will get damaged.
The charge controller maximum battery voltage might change, this depend on the design and manufacturer of the charge controller, always read your manual.

7.6.2 For PWM charge controller: same thing must be done.

Take note of the following;
1. Maximum PV voltages of the charge controller
2. Watts ratting of the charge controller
3. Battery bank voltage

Now, to size the charge controller to be used, in respect to above charge controller parameters;

i. We know our input PV voltage to be 83.26VDC
ii. We know our PV watts (total PV connected in series and parallel) to be 8,641.12 W
iii. Battery bank voltage to be 48VDC.

By using this simple equation, **Watts ÷ Volts = Amps**. You take the total watts of the solar array divided by the voltage of the battery bank. That will give you the output current of the charge controller.

For example, **8,641.12 W** solar array ÷ **48VDC** battery bank = **180 A.**

The rating of the charge controller should be at least 200 A.

We can decide to use 180A MPPT charge controller because some 180A MPPT charge controllers support up to 9,200W solar array at 48VDC battery voltage. But for the sake of safety and good efficiency, we will go for 200A MPPT solar charge controller.

If we use the charge controller with watts ratting given as follows;

- **For 180A MPPT maximum watts at 12VDC 2160W/ 24VDC 4320W/ 36VDC 6480W/ 48VDC 8640W**

> *"But, will result to poor efficiency of the charge controller. Because we have consumed total current capacity of the controller, any short circuit, electric surge, thunder strike can damage the charge controller". i.e. 180A controller capacity = 180A consumed".*

- **For 200A MPPT maximum watts at 12VDC 2400W/ 24VDC 4800W/ 36VDC 7200W/ 48VDC 9600W**

> *"Better efficiency, it can withstand any short circuit, electric surge, thunder strike without damage". i.e. 200A controller capacity = 180A consumed."*

So, the size of solar charge controller to be used is 200A and above.

Note: always use charge a controller whose current rating is higher than the required current. E.g. for a work that requires 60A an 80A charge controller should be used, because of safety and better efficiency of the charge controller.

7.6.3 *Charge* controllers connected in parallel

A solar charge controller can be connected in parallel to each other, just as the inverter has a parallel connection, the charge controller can also be connected in parallel.

For example, we are about to use a 200A charge controller, what if the engineer can't get a 200A charge controller in his own location will the work stop..?

The answer is NO..!

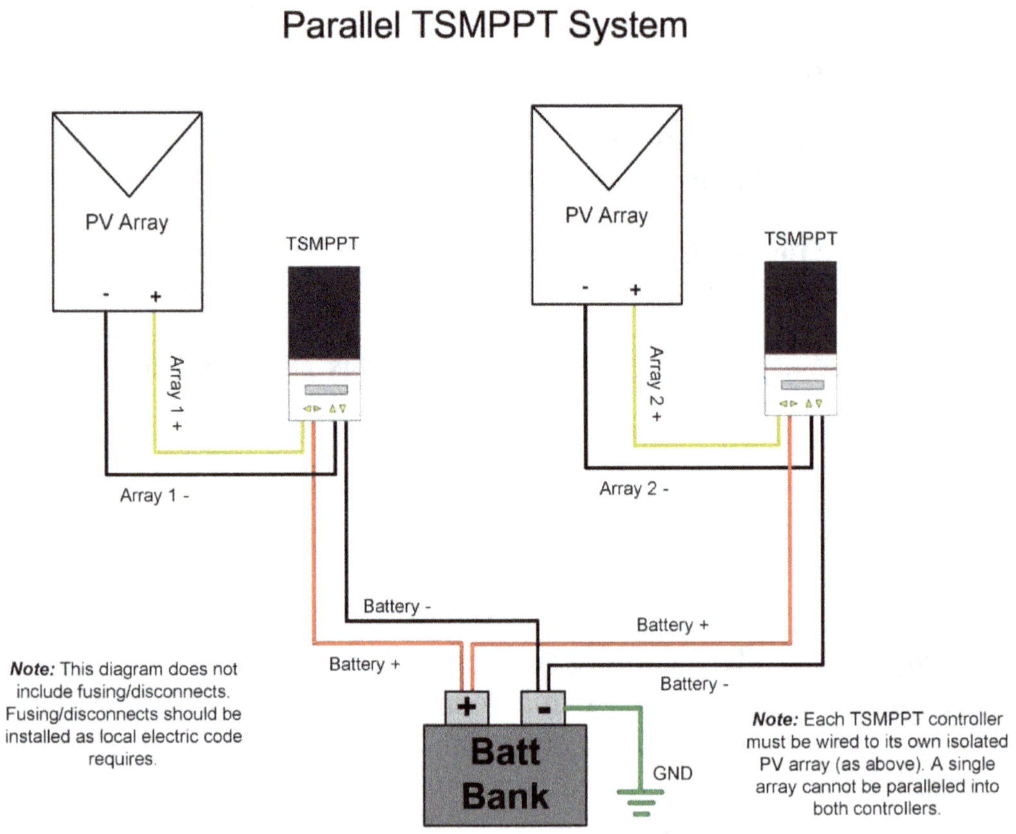

Parallel TSMPPT System

Figure 7.19 Two controllers configured in parallel

So, let's assume that the two charge controllers above are 100A each. To connect this charge controllers in parallel we must reconfigure our solar panel connections.
We have 16 solar panels, and the requirements of a 100A charge controllers is given below;

- For 100A MPPT maximum watts at 12VDC 1400W/ 24VDC 2800W/ 36VDC 3600W/ 48VDC 5400W
- For 100A MPPT maximum PV voltage is 200VDC
- For 100A MPPT battery voltage is 12VDC/24VDC/36VDC/48VDC

And we are working on a 48VDC battery bank, the maximum power voltage is given as 41.64VDC, by dividing the 16 solar panels into two strings I.e. 16 divided by 2 = 8.

Now we have 8 solar panels for each strings.

- **For string 1**: Eight (8) solar panels for each strings, two solar panels will be connected in series which will give us 83.28VDC, which is higher than the battery voltage 48VDC and lower than the maximum PV voltage 150VDC. By connecting two solar panels in series the PV power output becomes 1,080.14W and the other remaining 6 solar panels when connected in series and parallel which will give us a total power 4,320.56 the total power output is below the maximum solar input power for 100A charge controller which is 5,400W.

The PV connection for string one (1) is shown below.

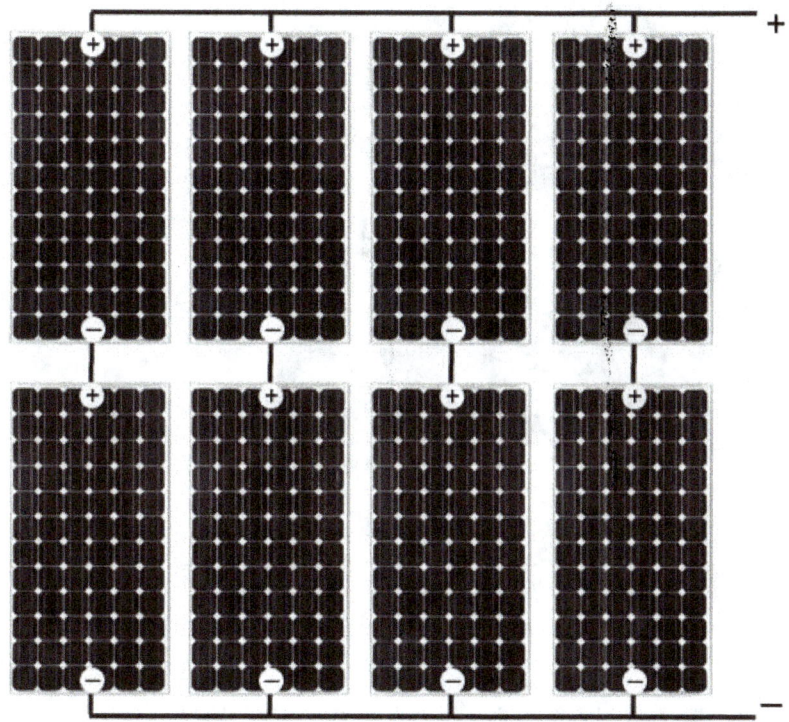

Figure 7.20 Series and parallel configuration for string one (1)

- **For string 2** Eight (8) solar panels for each strings, two solar panels will be connected in series which will give us 83.28VDC, which is higher than the battery voltage 48VDC and lower than the maximum PV voltage 150VDC. By connecting two solar panels in series the PV power output becomes 1,080.14W and the other remaining 6 solar panels when connected in series and parallel which will give us a total power 4,320.56 the total power output is below the maximum solar input power for 100A charge controller which is 5,400W.

The PV connection for string two (2) is shown below.

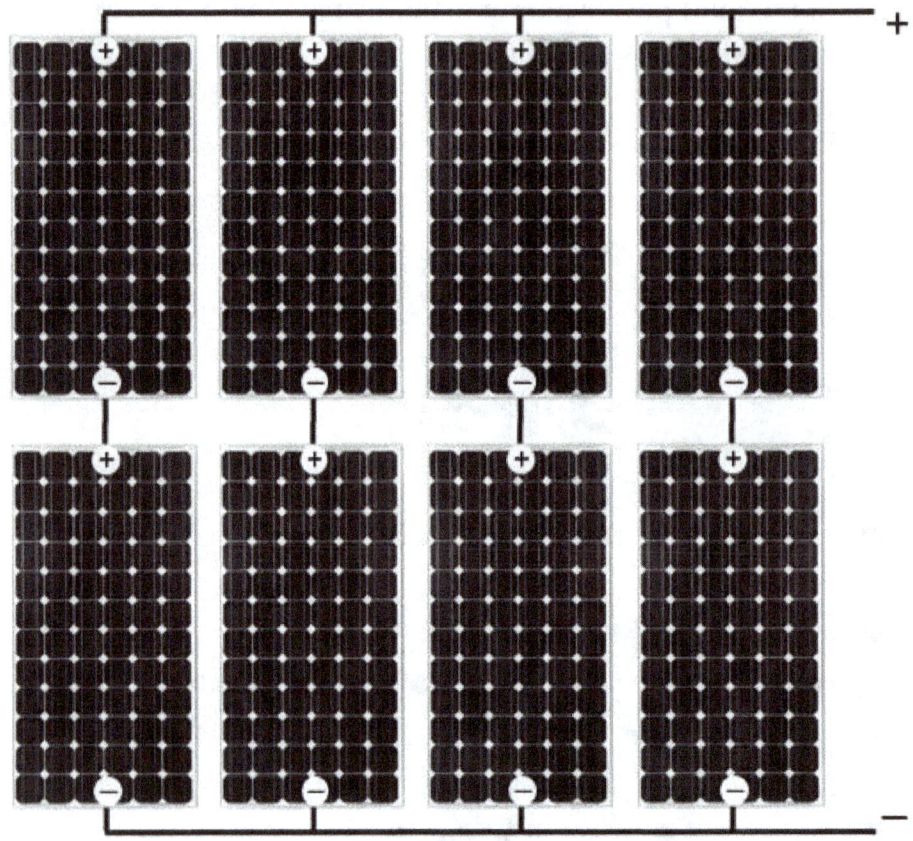

Figure 7.21 Series and parallel configuration for string two (2).

By adding the power output of both strings will give us **8,641.12W.** This tells us that if we use two charge controllers 100A connected in parallel or we use single charge controller 200A we will arrive at the same results.

Now, string one (1) solar array will be connected to charge controller one (1) also string two (2) solar array will be connected to charge controller two (2). As shown below.

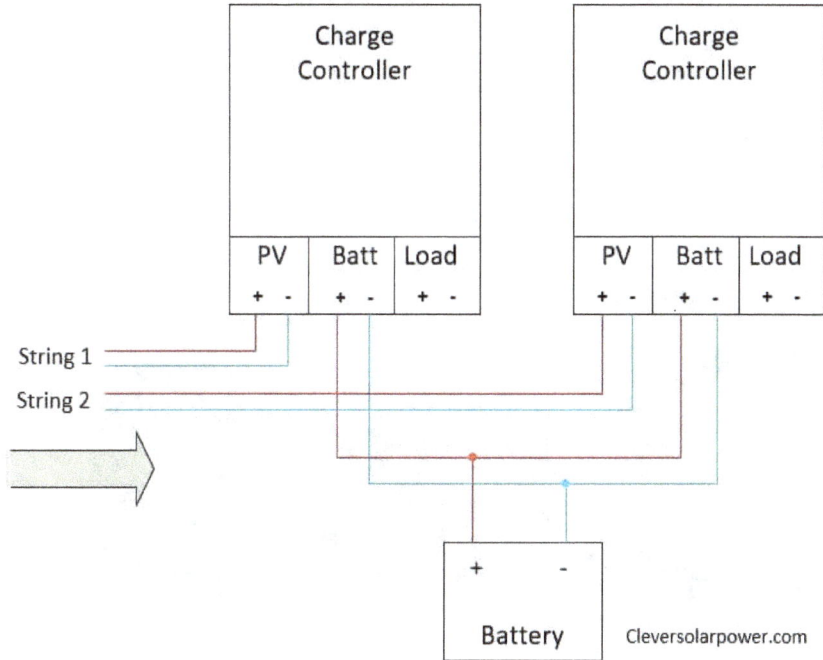

Figure 7.22 Two array strings in parallel

Note: only the battery output of the charge controllers can be parallel, do not

Parallel the solar array of string one (1) and string two (2) together.

CHAPTETR EIGHT

8.1. Photovoltaic System Application

1. Agricultural and livestock farms:

PV-powered pumping system

PV-powered solar dryer

PV-powered dairy system

PV-powered greenhouse

PV-powered crop protection system

Figure 8.1 Agricultural and livestock application of solar panel

Photovoltaic systems can also be used on farms. Farms that are far from power distribution lines. Photovoltaic are used to supply electricity at the farm which are used for lighting, motors, shearing machines, pumps etc. In livestock applications, solar photovoltaic systems are used to power pumps to provide water for livestock troughs. On specific farms, solar energy is used to power milking systems and milk cooling. Also even these systems are practical for electric fences.

1. **Irrigation controls**: Another use is to supply power for irrigation controls and solenoid valves. This technique has allowed a better distribution and saving of water, particularly in systems based on drip irrigation or low pressure.

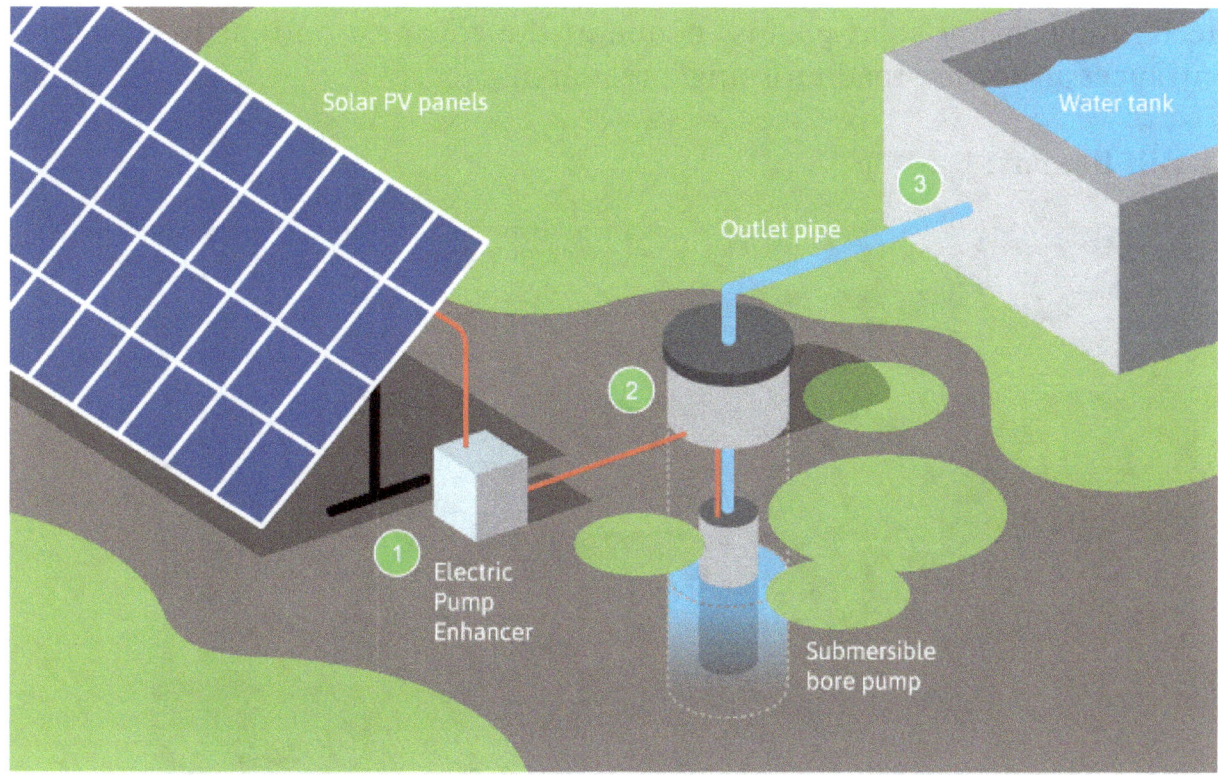

Figure 8.2 Irrigation control

2. **Street light:** photovoltaic systems are also used to power modern street lights, for visibility when there is power outage.

It is commonly used in remote areas with no access to electricity.

Figure 8.3 Solar Street light

Public lighting, using solar panels, is one of the cheapest options to illuminate the entrances in towns, road crossings, rest areas, etc.

3. **Bill boards:**

Figure 8.4 Bill board advertisement

This type of photovoltaic application can also be found in billboards, bus stops, street lighting, and tunnel lighting, among other examples.

4. Remote areas:

Figure 8.5 Solar application in remote areas

One of the essential applications of photovoltaic cells today is the power supply of small rural areas with a centralized system. Power in remote areas currently has all the comforts that can be had in a conventional electrical system. In addition, this system allows any appliance to replace fossil fuel dependency.

5. Signal elements:

Figure 8.6 Marine signal lights

Photovoltaic solar energy allows the automation of lighthouses and buoys for maritime use.

For aerial use, panels are being used to power beacons and signaling signs on the runways.

Another great use of solar cells is signaling roundabouts, curves, traffic signs, obstacles, etc., using high brightness LEDs. The low consumption of the LEDs allows a small PV panels to be carried out in these systems.

6. **Electric cars and electric charging stations**: Electric cars and trucks embedded with photovoltaic cells can convert energy from sunlight into electricity. Storing solar energy in batteries enables them to run smoothly at night or in the absence of direct sunlight.

Figure 8.7 Electric car

Solar panels can be attached to vehicle bodies using mechanical fasteners or structural adhesives. However, to be aerodynamic and aesthetically pleasing, automotive engineers prefer to integrate solar modules into body panels.

Figure 8.8 Electric charging station

7. Power in space:

From the beginning, PV has been a primary power source for Earth-orbiting satellites. High-efficiency PV has supplied power for ventures such as the International Space Station and surface rovers on the Moon and Mars, and it will continue to be an integral part of space and planetary exploration.

Figure 8.9 Solar space station

8. **Military Uses:**

Lightweight, flexible thin-film PV can serve applications in which portability or ruggedness are critical. Soldiers can carry lightweight PV for charging electronic equipment in the field or at remote bases.

Figure 8.10 Military use of solar panel

Solar-powered operations also allow military units to operate more stealthily. Unlike the noisy diesel generators that are usually used to power lights and other equipment in the field, solar panels don't produce any noise. Their silent operation reduces the likelihood of detection by enemy forces.

9. **Solar lighting:**

Solar lights have become commonplace. You can find them everywhere from home landscaping and security lights to road signs and street lights. These solar lighting technologies are inexpensive and readily available.

Figure 8.11 Solar lighting

CHAPTAER NINE

9.1. Solar Panel and battery maintenance

Although solar panels require some maintenance compared to other power systems, you should periodically perform a few simple maintenance tasks.

Figure 9.1 Solar panel maintenance

9.1.1. Solar panel maintenance

1. Check the panels for dust, if the system is in a dusty location with little rain, the panels may need to be cleaned periodically. Clean the modules with water and mild soap. Avoid solvents or strong detergents.

Figure 9.2 Washing solar panel with water

2. Check to see if there are any shade problems due to vegetation or a new building. If there are, make arrangements for removing the vegetation or moving the panels to a shade-free place.

Figure 9.3 Shaded solar panel

3. Check the panel mounting to make sure that it is strong and well attached. If it is broken or loose, repair it.

Figure 9.4 Checking for broken frames

4. Check that the glass of the panels is not broken. If it is, the panel will have to be replaced.

Figure 9.5 Broken solar panel

5. The combiner box should be checked periodically for weather protection, tightness of wires and water seals.

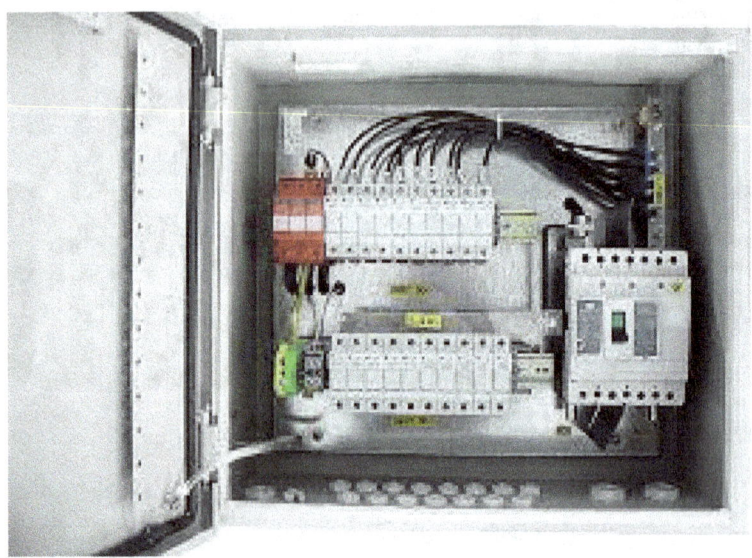

Figure 9.6 Photovoltaic combiner box

9.1.2. Batteries maintenance

Figure 9.7 Battery maintenance

9.1.2.1 Dry cell batteries

Battery maintenance depends largely on battery type, though all batteries require periodic inspection to verify system operation.

1. Check connections for tightness and corrosion. Clean and tighten as needed.

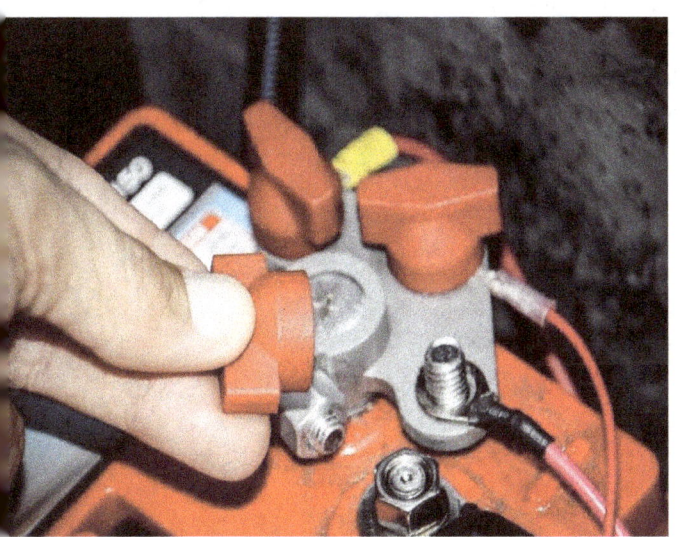 

Figure 9.8 Loosen battery terminal and corroded batter terminal

2. Cover connections with heavy grease. Do not get the grease on any part of the battery except the connections.

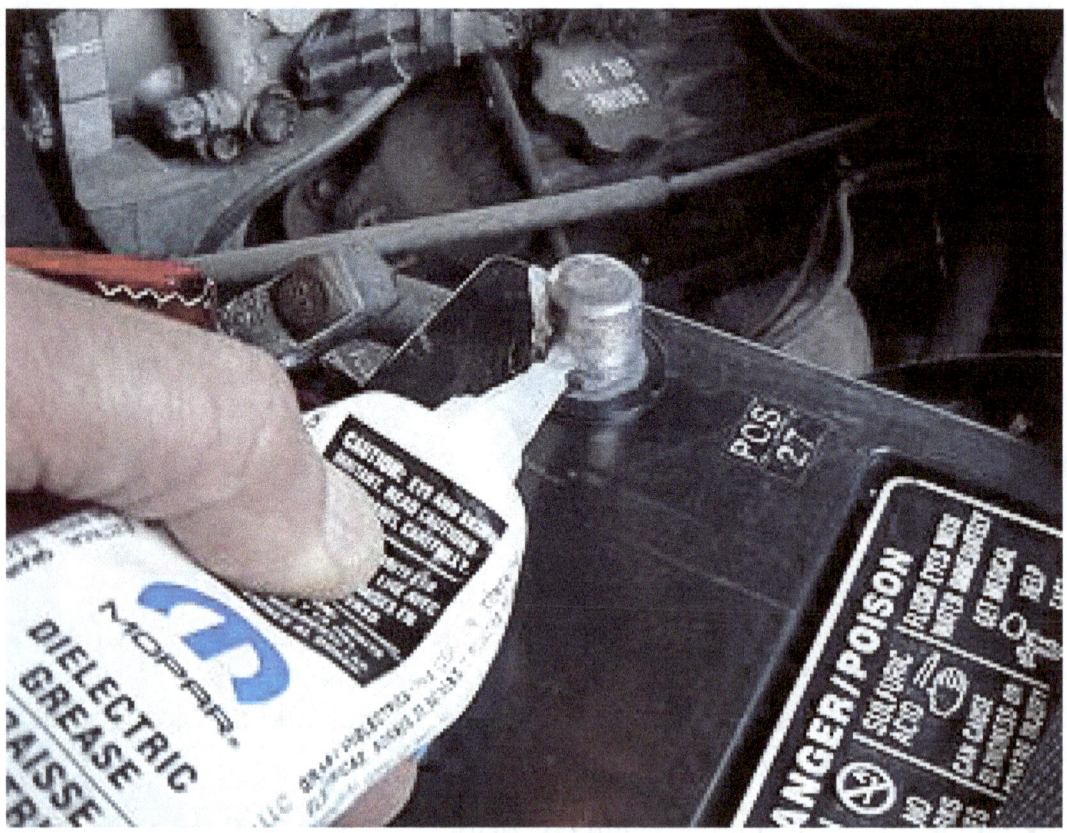

Figure 9.9 Greasing a battery terminal

3. Clean the battery with fresh water and a rag. The acid and the corrosion on the battery top should be washed off with the cloth moistened with baking soda or ammonia and water.

Figure 9.10 Cleaning a battery terminal with cloth

9.1.2.2 Lead-acid batteries

1. Clean the top of the battery. Check connections for tightness and corrosion. Clean and tighten as needed.

Figure 9.11 Loosen battery terminal and corroded batter terminal

2. Check each cell with a hydrometer and record the readings. When checking, take off one cap at a time. Do not remove all caps at once because that greatly increases the risk of dirt getting into the cells.

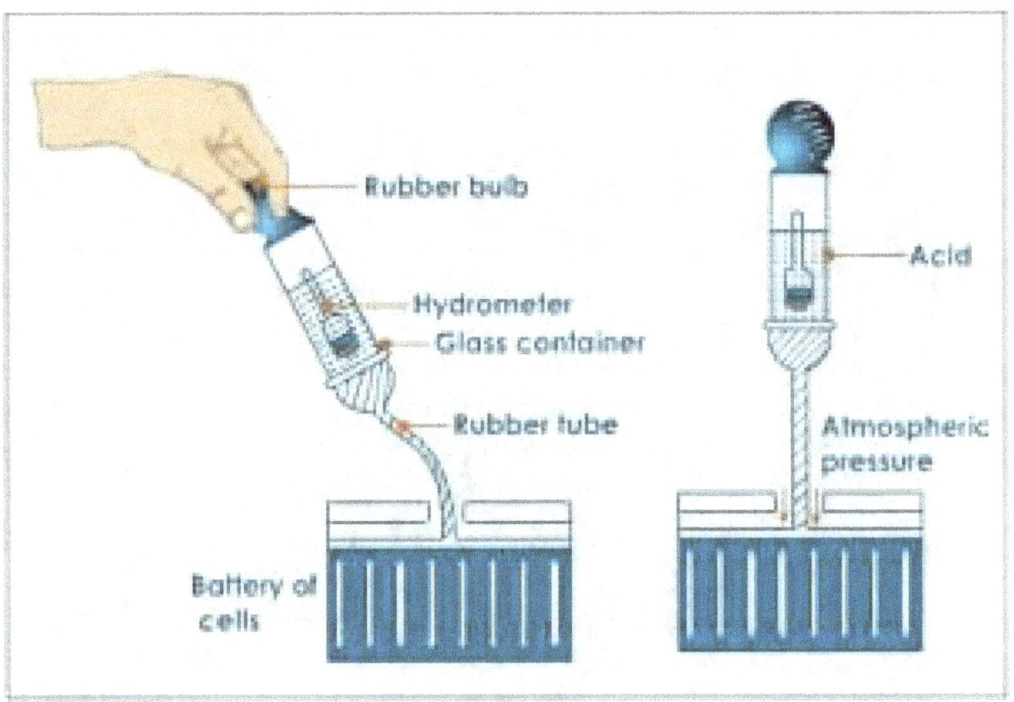

Figure 9.12 Using hydrometer to check water level

The level of the electrolyte should always be 10 to 15 mm above the top of the plates. If any cells are low on water, add distilled water to obtain to the correct level. Never add more acid, only water. If distilled water is not available, carefully collected rainwater can be used. Remember that any salt, minerals or oil in the water will poison the battery and shorten its life, so be very careful about collection and storage of water for the battery.

3. If any of the water level indicator of the battery have been lost or broken, cover the fill holes loosely with plastic or glass until you get a new water level indicator.

Figure 9.13 Water level indicator

Never cover the holes with paper, cork, cloth or metal. Never leave the holes uncovered. Be careful that the temporary cover that you install does not block the holes tightly because the cells must have air.

4. Clean the battery with fresh water and a rag. The acid and the corrosion on the battery top should be washed off with the cloth moistened with baking soda or ammonia and water.

Figure 9.14 cleaning a battery terminal with cloth

5. When the water level has dropped below the minimum on the water level indicator, refill the battery with a distilled water till the water level indicator returns back to maximum.

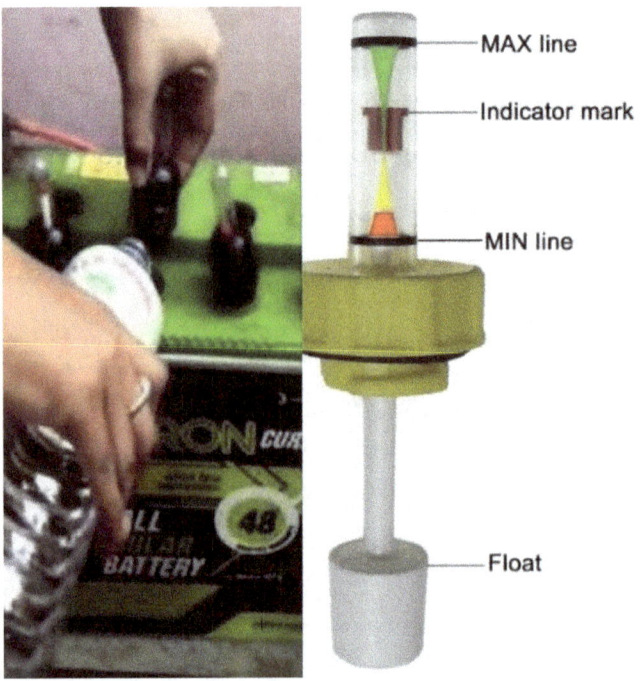

Figure 9.15 Refilling battery with distilled water

9.2. Troubleshooting an inverter

Figure 9.16 troubleshooting an inverter

COMMON INVERTER PROBLEMS	EXPLANATION/ POSSIBLE CAUSES	POSSIBLE SOLUTION
over load	• The inverter is overloaded, with too many appliances on it. • Short circuit condition	• Try to turn off some loads on the inverter. • Check if there is no bridge between the neutral and live cable. • If problem continues after switching off high watt ratting appliances, take inverter out for repair.

fan not working but, inverter is on	When the cooling fans are not working, it can lead to overheating of the Inverter and cause major damages to the inverter that is beyond repair.	• uninstall the inverter • Allow the capacitors to discharge current, if not you might get shocked. • Try to fix the fan wires properly on the mother board. • If fans are still not working. replace the fan • if problem exist, take inverter out for repair
Over Temperature	If the internal temperature of the inverter is too high. Inverter temperature ranging from 70 degree Celsius to 100 degree Celsius.	• Open the inverter and allow the inverter capacitors to discharge correct, if not you might get shocked. • Check if the cooling fans are working, check if the fan cables are properly connected to the mother board. • Check if the thermistors cables are properly connected to the mother board. • Check if the air flow of the fan is not blocked by dust and particles. If yes clean properly. • If fans are not working, change the fan. • If thermistors are not working, change the thermistor. • If problem continues, take inverter out for repair.
Inverter always showing battery low	• Battery voltage too low. • Battery need to be recharged.	• Turn on the grid power to recharge the battery. • During the day, check if the PV circuit breaker is not turned OFF, if yes turn ON the PV circuit breaker.

		- Use a digital meter to check the amount of voltage coming from the PV array. And also the battery voltage.
- Carry out continuity test to know if current flows from the charge controller to the battery. Also check if there is continuity from the grid source circuit breaker to the inverter input
- If charge controller is freeze, restart the charge controller.
- If problem continues, disconnect the battery from inverter and boost charge the battery by directing the solar array on the battery bank for some minutes. Or use a battery charger to boost charge.
- If problem continues, replace the battery.
System Shut Down	This can happen if the battery voltage is too low, and no PV array connected	- Recharge the battery.
- Carry out all the troubleshooting steps above.
- Replace battery.
Inverter Is Showing AC Input Voltage, When Ac Input Source Is Not Available	- A load (appliance) returns current back to the inverter.
- Improper load isolation at the distribution box.
- Short circuit.	- Turn off the grid source from the changeover switch and Use an electric current tester to check if there is a current at the input terminal of the inverter both live and neutral cable. If yes, remove both input cables.
- Open the electric distribution box and carry out a proper load isolation.
- Check for a damaged appliances, power transfer switches etc.

		that can return current back to the inverter through the neutral cable
The inverter is making a humming sound	• Return current from inverter output to the input. • High input voltage • Bad input frequency	• Check the input voltage. • Check the input frequency. • Turn off the grid source from the changeover switch and Use an electric current tester to check if there is a current at the input terminal of the inverter both live and neutral cable. If yes, remove both input cables. And carry out a proper load isolation. • Open the electric distribution box and carry out a proper load isolation. • Check for a damaged appliances such as bulb, fan, power transfer switches etc. that can return current back to the inverter through the neutral cable • If problem continues, take inverter out for repair.
parallel input phase error	Wrong input phase connection	• Check the AC input connection of the inverter. • Check the inverters input breakers and ensure it is ON • Check the communication cable is properly connected. • For split phase and single phase configuration. • Check if the current sharing cables are properly connected

		for single phase parallel configuration. • If problem continues take inverter out for repair.

Figure 9.16 Fault finding on circuit breakers

Battery overcharged	• The battery voltage is too high. • Battery is over charged	• Check the battery voltage. • Ensure that the charge controller is regulating the PV voltage, direct connection PV to the battery when the charge controller is damaged, can make the inverter to detect a false battery voltage and write "battery overcharged" on the screen. • Check battery quality

			• If problem continues, take inverter out for repair.
Inverter Is Shocking		• Wrong wiring at either the input or output terminal • Damaged internal component such as relay. • Short circuit.	• Check if the wires are properly fitted at the input and output terminal. • Check if the EARTH port is not looped to LIVE (HOT) port. Either at the input terminal or output terminal of the inverter. • Take inverter out for a repair. There is a damaged component inside the inverter.
Inverter Is On But No Output		• Output fuse blown. • Either the output live or neutral cable is removed • Tripped output breaker.	• Check if the output breaker of the inverter is ON. If it has tripped turn in ON. • Check the live and neutral wires at the output terminal of the inverter are correctly connected. If reversed make corrections. • Remove the output fuse of the inverter and carry out a continuity test on it. If there is no continuity on the fuse, the fuse is blown. Replace the fuse with a new fuse and the same current ratting.
Inverter Shut Down Completely During Start Up Process		• Low Battery • Failed internal components	• Check if the battery voltage. • Check if the battery polarity are not reversed. • Take inverter out for repair
No Response After Power On		• Battery polarity reversed. • battery low	• Check if the battery polarities are not reversed. • Recharge the battery and restart the inverter.

Grid Source Is Available But Inverter Still In Battery Mode	• The inverter does not by-pass the AC input. • If AC input is zero on the screen. AC input Circuit breaker must have tripped. SOLAR, BATTERY, UTILITY priority settings are not properly set.	• Check the AC input connections. • If the AC input is zero, check the AC input circuit breaker, if the breaker has tripped turn it ON. • Check if the solar, battery and utility priority settings are correct • If the by-pass switch of the inverter is switched off, turn it on, and also put the inverter on UPS mode.
Inverter Not Charging	This can happen for different reasons. • The AC input voltage is too high or too low • The AC input frequency is too high or too low. • Fluctuating input AC voltage • The PV arrays are disconnected. • Tripped AC input breaker	• Check the AC input connections. • Check the AC input circuit breaker if it has tripped or not. • Use a digital meter to check the input AC voltage to know if it is too high or too low. If the AC input voltage is too high, the rectifier circuit will not accept it for battery to charge. Reduce the input voltage either with a voltage regulator or a step down transformer. • If the input frequency is too high or too low, buy a stabilizer that its capacity is higher than the inverter capacity. To stabilize the frequency. • For fluctuating AC input voltage, get a stabilizer. • If the AC input source is from a generator, check its output voltage and frequency, if it is too high, call a technician to regulate the generator output voltage and frequency. • Check if the PV circuit breaker is not turned off.

ELECTRIFIED

| When Switched On Inverter Relays Are Switching On And Off Repeatedly | • Battery is disconnected
• Reversed polarity. | • Check the battery connection properly.
• Check the polarity of the battery if properly connected.
• If problem continues, take inverter out for repair. |

Table 9.1 troubleshooting an inverter

www.ingramcontent.com/pod-product-compliance
Lightning Source LLC
Chambersburg PA
CBHW080912170526
45158CB00008B/2078